Duong-Van Nguyen

Perceptual Inference for Autonomous Navigation

AF092439

Duong-Van Nguyen

Perceptual Inference for Autonomous Navigation

Vegetation Detection and Terrain Classification

Südwestdeutscher Verlag für Hochschulschriften

Impressum / Imprint
Bibliografische Information der Deutschen Nationalbibliothek: Die Deutsche Nationalbibliothek verzeichnet diese Publikation in der Deutschen Nationalbibliografie; detaillierte bibliografische Daten sind im Internet über http://dnb.d-nb.de abrufbar.
Alle in diesem Buch genannten Marken und Produktnamen unterliegen warenzeichen-, marken- oder patentrechtlichem Schutz bzw. sind Warenzeichen oder eingetragene Warenzeichen der jeweiligen Inhaber. Die Wiedergabe von Marken, Produktnamen, Gebrauchsnamen, Handelsnamen, Warenbezeichnungen u.s.w. in diesem Werk berechtigt auch ohne besondere Kennzeichnung nicht zu der Annahme, dass solche Namen im Sinne der Warenzeichen- und Markenschutzgesetzgebung als frei zu betrachten wären und daher von jedermann benutzt werden dürften.

Bibliographic information published by the Deutsche Nationalbibliothek: The Deutsche Nationalbibliothek lists this publication in the Deutsche Nationalbibliografie; detailed bibliographic data are available in the Internet at http://dnb.d-nb.de.
Any brand names and product names mentioned in this book are subject to trademark, brand or patent protection and are trademarks or registered trademarks of their respective holders. The use of brand names, product names, common names, trade names, product descriptions etc. even without a particular marking in this work is in no way to be construed to mean that such names may be regarded as unrestricted in respect of trademark and brand protection legislation and could thus be used by anyone.

Coverbild / Cover image: www.ingimage.com

Verlag / Publisher:
Südwestdeutscher Verlag für Hochschulschriften
ist ein Imprint der / is a trademark of
OmniScriptum GmbH & Co. KG
Heinrich-Böcking-Str. 6-8, 66121 Saarbrücken, Deutschland / Germany
Email: info@svh-verlag.de

Herstellung: siehe letzte Seite /
Printed at: see last page
ISBN: 978-3-8381-3988-3

Zugl. / Approved by: Muster: Siegen Universität, Diss., 2013

Copyright © 2014 OmniScriptum GmbH & Co. KG
Alle Rechte vorbehalten. / All rights reserved. Saarbrücken 2014

This dissertation is dedicated to my parents, my wife and my sister.

Abstrakt

Diese Arbeit beleuchtet sieben neuartige Ansätze aus zwei Bereichen der maschinellen Wahrnehmung: Erkennung von Vegetation und Klassifizierung von Gelände. Diese Elemente bilden den Kern eines jeden Steuerungssystems für effiziente, autonome Navigation im Außenbereich.

Bezüglich der Vegetationserkennung, wird zuerst ein auf Indizierung basierender Ansatz beschrieben (1), der die reflektierenden und absorbierenden Eigenschaften von Pflanzen im Bezug auf sichtbares und nah-infrarotes Licht auswertet. Zweitens wird eine Fusionmethode von 2D/3D Merkmalen untersucht (2), die das menschliche System der Vegetationserkennung nachbildet. Zusätzlich wird ein integriertes System vorgeschlagen (3), welches die visuelle Wahrnehmung mit multi-spektralen Methoden kombiniert. Aufbauend auf detaillierten Studien zu Farb- und Textureigenschaften von Vegetation wird ein adaptiver selbstlernender Algorithmus eingeführt der robust und schnell Pflanzen (bewuchs) erkennt (4). Komplettiert wird die Vegetationserkennung durch einen Algorithmus zur Befahrbarkeitseinschätzung von Vegetation, der die Verformbarkeit von Pflanzen erkennt. Je leichter sich Pflanzen bewegen lassen, umso größer ist ihre Befahrbarkeit.

Bezüglich der Geländeklassifizierung wird eine struktur-basierte Methode vorgestellt (6), welche die 3D Strukturdaten einer Umgebung durch die statistische Analyse lokaler Punkte von LiDAR Daten unterstützt. Zuletzt wird eine auf Klassifizierung basierende Methode (7) beschrieben, die LiDAR und Kamera-Daten kombiniert, um eine 3D Szene zu rekonstruieren.

Basierend auf den Vorteilen der vorgestellten Algorithmen im Bezug auf die maschinelle Wahrnehmung, hoffen wir, dass diese Arbeit als Ausgangspunkt für weitere Entwicklung en von zuverlässigen Erkennungsmethoden dient.

Abstract

Environment sensing is required in order for robots to operate safely in either shared workspaces between robot and human or unpredictable natural environments. However, available perceptual inference algorithms require many smoothness assumptions such as a flat ground plane, straight walls, and so on; thus their efficiency depends on the degree of smoothness of the beliefs. In the real world, such these assumptions often fails, leading to unreliable perceptual inference results. In fact, there exists some investigations on making perceptual inference robust, but the results vary significantly under different outdoor scenarios. This is caused by the lack of information due to the range discontinuities given by a LiDAR. Hence, robustly classifying terrains into object types which benefit autonomous navigation is still a challenging problem. Alternatively, current autonomous navigation techniques only work well in highly structured environments such as on-road, hallway and campus where objects are usually rigid and static, but fail to deal with cluttered outdoor environments. Particularly, vegetated terrain introduces one more degree of freedom to the problem that what is considered as an "obstacle" from a purely geometric point of view, may not represent a danger for the vehicle if it is composed of compressible vegetation. While current perception-based techniques do not operate efficiently in terrains containing vegetation, the most reliable way for navigation in such situations is to detect vegetation areas in the viewed scene, and then enable possible strategies to cope up with.

Motivated by concrete robotics problems, we explicitly pursue solutions for two perception tasks: Vegetation Detection and Terrain Classification, which are at the core of any control system for efficient autonomous navigation in outdoor environments. Consequently, we have achieved eights contributions related to those tasks, which have been published in peer-review journals and conferences, and are also introduced in this thesis.

Regarding vegetation detection, we first describe a vegetation indices-based method (1), which relies on the absorption and reflectance properties of vegetation to visual light and near-infrared (NIR) light, respectively. Unlike previous art which focused on applying polarized filters and colour filters to reduce illumination effects, we study a new multi-spectral device which is equipped an active NIR lighting system. By adding such independent light, the NIR reflectance is stabilised by adjusting the light intensity, and then a stable multi-spectral system is achieved by simply setting the gain of colour sensor as an off-set of that of NIR sensor. We will show through practical experiments that the proposed system setup really provides the most stable multi-spectral system available. Within using the proposed multi-spectral system, we devise a new vegetation index, the so-called Modification of Normalized Difference Vegetation Index, through a regression analysis on red and NIR reflectance changes in term of luminance. Through evaluation on a diverse set of databases given by real robotics experiments, it is confirmed that the new index far outperforms other indices as well as other available methods with regard to vegetation detection in different lighting conditions, and under different illumination effects. Since vegetation detection is very easy done by human eye, human perception-based methods are also of our interest. Within this regard, we present a 2D/3D feature fusion method (2), which collects the world information from a CMOS camera and a LiDAR in order to extract and fuse 2D/3D features to generate vegetation classifier. The classifier performs well and far better than previous classification-based methods in accuracy, but similar in processing time. On the other hand, a general vegetation detection using an integrated vision system (3) is proposed to realise our greedy ambition in combining visual perception-based and multi-spectral methods by only using a single device. The device is termed MultiCam, which mounts both the CMOS sensor and the Photo Mixer Device (PMD) sensor into a molecular setup, and thus provides simultaneously colour, NIR intensity and depth information. Even though the given depth information is not accurate enough for any geometric distribution analyses, there are still good spatial features extracted based on the interesting finding that there is a significant reduction of illumination noise inside vegetation regions in the given depth image; this may be explained by the strong reflectance of vegetation to the NIR light. Since the MultiCam can operate as fast as a regular video camera, this approach is able to capture both visual and spectral reflectance properties of vegetation while still producing high frame rate. Consequently, this approach provides much higher accuracy and higher frame rate than previous classification-based approaches or even the 2D/3D feature fusion approach. By observing that the method (1) can be comparable to the method (3) in accuracy in most cases, except it could not distinguish between vegetation and warm or strong NIR reflectance objects and its performance in dim lighting conditions is poor; in contrast, the method (1) produces the frame rate as six times higher as that of the method (3). We come out with the idea of creating a fast adaptive learning algorithm to detect general vegetation with the follow-up of the system setup in the method (1). The algorithm is termed Spreading Algorithm (4). It is an iterative region-growing technique coupled with annealing, which is based on an annealed criteria, a convex combination of colour and texture dissimilarities. Remarkably, the unstructured texture feature derived within this work is really distinct, which is able to intuitively distinguish vegetation from other artificial dense edge objects; this is infeasible in previous approaches. Instead of building colour models, this method conducts an intensity shift-invariant colour feature which is able to guarantee detecting variety of vegetation appeared in different colours. Overall, the Spreading Algorithm far outperforms the state-of-the-art, or provides the most efficient and robust vegetation detection mechanism. Finally, in order to answer the question if the detected vegetation is passable or not, we present an active approach for a double-check of passable vegetation detection (5). The novel approach relies on the compressibility or less-resistance of vegetation, which is supposed to be movable by a strong wind. Hence, we design our robot system with blowing devices which create strong wind to effect vegetation. Motion detection and motion compensation techniques are applied detect moving objects. Moving vegetation is pointed out through a mapping between the moving objects and the detected vegetation (using one of the above methods). On the other hand, the degree of resistance of vegetation is estimated by recording its movement through an optical flow process. Consequently, the lower degree of resistance vegetation has, the more traversable it is. For the purpose of autonomous navigation, the region of interest should be right at the front of the robot. Within that restriction of region of interest, the performance of the method is really impressive with high stability, accuracy and efficiency, which has been confirmed through many real robotics experiments in both morning and afternoon conditions.

Regarding terrain classification, we first introduce a structure-based method (6) to capture the world scene by inferring its 3D structures (linear, scatter, surface) through a local point statistic analysis. Instead of sliding a cube in space to select a local region to be analysed, a segmentation of the point cloud in terms of homogeneous distance and neighbourhood is proposed to result objects in form of regions of interest. Thus, the proposed method is more flexible to cope with different terrains, and avoids the problematic selection of the cube's size. As a result, the classification accuracy is improved about 10% in average. Furthermore, this method conducts a surface smoothness estimation through measuring distance variation inside edgeless regions, thus it is able to classify rough surface objects (low grass, bushes) and smooth surface objects (wall, concrete road). Secondly, due to the lack of objective information because of range discontinuities given by LiDAR data, we propose a novel approach (7) which combines the LiDAR data and colour information to reconstruct a 3D scene. Consequently, object representation is described more details, thus enabling an ability to classify more object types including tree trunk, human, wall/building, vegetation, sky, and road; this is infeasible in previous approaches.

Based on the success of the proposed perceptual inference methods in the environmental sensing tasks, we hope that this thesis will really serve as a key point for further development of highly reliable perceptual inference methods.

Acknowledgements

This work would be incomplete without acknowledging the many people who made this thesis possible. In fact, the number of people I need to thank will not fit to a single Acknowledgement section, I would like to spend this opportunity to express my profound gratitude to some whose contribution is obvious.

First and foremost, my most sincere gratefulness must go to my supervisor, Prof. Dr. Klaus-Dieter Kuhnert, for his consistent help, invaluable academic guidance and attention during the whole work. His engagement, scientific knowledge, encouragement and continuous support during the past years were crucial not only for the accomplishment of this work, but also for the expansion of my scientific knowledge and my growing interest in the world of Robotics.

I would like to give my thanks and appreciation to Prof. Dr. Volker Blanz for his role as the second supervisor. He advised and conducted me to a professional and scientific manner in presenting this work.

I am deeply grateful toward Prof. Dr. Otmar Loffeld who has not only given me valuable advices and suggestions to improve this work, but also inspired and motivated me by his enthusiasm for his work. Still, I would not forget how he lighted up my world by giving me the opportunity to join the Research School MOSES.

I would like to thank Dr. Wolfgang Weihs for his valuable suggestions and fruitful discussions related to Time-of-Flight sensors.

I wish to thank Dr. Holger Nies for his great work as the managing director of the IPP/MOSES program, who is willing to give any necessary help. Without the detailed documentary support from Mrs. Niet-Wunram Silva and Mrs. Waltraud Setzer, completing administrative procedure transaction would have been much more difficult and time-consuming.

I would also like to give a big thank to Stefan Thamke for helping me in correcting the German version of my thesis summary.

Special thanks go to many dedicated co-workers at the AMOR lab because working on multiple sensors and devices mounted on a large outdoor robot with the size of a small car is really an extreme challenge, which would not be possible to take all the work done without their collaborative effort.

Special thanks are extended to my officemate Tao Jiang, who beared with me throughout the past four years. Thank you for being as a close friend the whole time and supporting me. In this context, I would also wish to thank my other colleagues Lars Kuhnert, Stefan Thamke, Markus Ax, Jens Sclemper, and Ievgen Smielik for the close collaboration in the projects, nice and intensive discussions and providing me with much technical support. Thank you all you guys in the AMOR team for your collaboration in writing publications and sharing useful knowledge in computer vision, robotics and autonomous systems.

Members of IPP/MOSES program also deserve my sincerest thanks, their friendship and assistance has meant more to me than I could ever express.

Last but not least, I want to express my gratitude to my parents and my sister, whose love and encouragement have supported me throughout my education. And finally, I am very thankful for the unwavering support received from my wife Phuong, who always accompanies with me and constantly provides comfort during difficult times when nothing seemed to work out right.

This thesis has been funded by the Research School MOSES at the Centre for Sensor Systems of University of Siegen. Their support is gratefully acknowledged.

Preface

The work outlined in this dissertation was carried out in the Research School on Multi-Modal Sensor Systems for Environmental Exploration, Centre for Sensor Systems, University of Siegen, over the period from April 2009 to April 2013. This dissertation is the result of my work and includes nothing which is the outcome of work done in collaboration, except for a few instances which are stated in the text. The material included in this thesis has not been submitted for a PhD degree or diploma or any other qualification at any other university. Furthermore, no part of my dissertation has already been or is currently submitted for any such degree, diploma or other qualification.

The dissertation presents a number of novel approaches for vegetation detection and terrain classification using different sensor systems under different configurations. Hence, it would be very large or might causes confusion when writing the thesis as a traditional monograph. While the number of publications is adicate for the thesis to be written as cumulative texts, the understanding of the whole work done within the PhD might be restricted under this format. I find the best way to present this PhD thesis in the **mixture** between **monograph** and **cumulative** format. Whereby, different approaches are divided into different groups with respect to their similarities, then each group is presented in one chapter. Additional chapters will be added to introduce the motivation and contributions of the thesis, as well as describe how concretely experiments carried out and the comparison between available approaches.

Contents

Contents		xvii
List of Figures		xxiii
List of Tables		xxxi
1 Introduction		**1**
1.1	Motivation	1
1.2	Problem Description	4
	1.2.1 Terrain Classification	4
	1.2.2 Vegetation Detection	5
1.3	Goald of this Thesis	7
1.4	Novel Contributions of the Thesis	8
	1.4.1 Fitting plane Algorithm-based Depth Correction for Tyzx DeepSea Stereoscopic Imaging	8
	1.4.2 Vegetation Indices Applied for Vegetation Detection	10
	1.4.3 2D/3D Feature Fusion for Vegetation Detection	12
	1.4.4 General Vegetation Detection Using an Integrated Vision System	15
	1.4.5 Spreading Algorithm for Efficient Vegetation Detection	17
	1.4.6 A Novel Approach for a Double-Check of Passable Vegetation Detection in Autonomous Ground Vehicles	20

CONTENTS

  1.4.7 Terrain Classification Based on Structure for Autonomous Navigation in Complex Environments 22
  1.4.8 A Novel Approach of Terrain Classification for Outdoor Automobile Navigation 24
 1.5 Document Structure 26
 1.6 Publications 27

2 Fundamentals 33
 2.1 The Experimental Platform AMOR 34
 2.2 Light Detection And Ranging (LiDAR) 36
  2.2.1 Optical Triangulation for 3D Digitizing 37
  2.2.2 Laser Pulse Time-of-flight 40
  2.2.3 Laser Phase-Shift Range Finder 41
  2.2.4 Laser Scanner SICK LMS221 42
 2.3 Structured Light 43
 2.4 The MultiCam 43
 2.5 Stereoscopic Imaging 46
  2.5.1 Fitting Plane Algorithm-based Depth Correction for Tyzx DeepSea Stereoscopic Imaging 49
   2.5.1.1 Introduction 50
   2.5.1.2 Planar Surface for Scene Understanding 53
   2.5.1.3 Fitting Plane Algorithm 58
   2.5.1.4 Experiments and Results 60
   2.5.1.5 Conclusion 62
 2.6 Multi-spectral Imaging 62

3 Vegetation Indices Applied for Vegetation Detection 65
 3.1 Related Work 69
  3.1.1 Ratio Vegetation Index 69
  3.1.2 Normalized Difference Vegetation Index 70
  3.1.3 Perpendicular Vegetation Index 70

	3.1.4 Difference Vegetation Index	71
	3.1.5 Soil-Adjusted Vegetation Index	72
	3.1.6 Modified Soil Adjusted Vegetation Index	72
3.2	A Novel Vegetation Index : Modification of Normalized Difference Vegetation Index	72
	3.2.1 Derivation of Novel Index	73
3.3	Experiments and Results	79
3.4	Conclusion	82

4 2D-3D Feature Fusion-based Vegetation Detection — 85
- 4.1 Related Work — 86
- 4.2 2D/3D Mapping — 87
- 4.3 3D point cloud analysis — 91
  - 4.3.1 Scatter Feature Extraction — 93
- 4.4 Colour Descriptors — 94
- 4.5 Support Vector Machine — 97
- 4.6 Experiments and Results — 99
- 4.7 Conclusion — 100

5 General Vegetation Detection Using an Integrated Vision System — 101
- 5.1 System Set-Up — 103
- 5.2 Spatial Features — 105
- 5.3 Vegetation Index Calculation — 107
- 5.4 Colour and Texture Descriptors — 109
- 5.5 Experiments and Results — 112
- 5.6 Conclusion — 115

6 Spreading Algorithm for Efficient Vegetation Detection — 117
- 6.1 Introduction — 118
- 6.2 Discussion on Vegetation Indices — 122
- 6.3 Visual Features for Scene Understanding — 123
  - 6.3.1 Opponent Color Space — 124

CONTENTS

		6.3.2	Unstructured Texture	125
	6.4	Spreading Algorithm		129
	6.5	Experiments and Results		133
	6.6	Conclusion		142

7 A Novel Approach for a Double-Check of Passable Vegetation Detection in Autonomous Ground Vehicles — 143

- 7.1 Introduction 144
- 7.2 Multi-spectral-based Vegetation Detection 146
  - 7.2.1 Standard Form of Vegetation Index 146
  - 7.2.2 Modification Form of Vegetation Index 147
  - 7.2.3 Convex Combination of Vegetation Indices 147
- 7.3 System Design 148
- 7.4 A Double-Check for Passable Vegetation Detection 151
- 7.5 Experiments and Results 155
- 7.6 Conclusions 158

8 Terrain Classification Based on Structure for Autonomous Navigation in Complex Environments — 159

- 8.1 Introduction 160
- 8.2 Methodology 163
  - 8.2.1 Efficient Graph-based Segmentation Technique ... 164
  - 8.2.2 Feature Extraction 165
    - 8.2.2.1 Neighbour Distance Variation Inside Edgeless Regions 165
    - 8.2.2.2 Conditional Local Point Statistics 168
  - 8.2.3 Support Vector Machine 169
- 8.3 Experiments and Results 171
- 8.4 Conclusion 173

9 A Novel Approach of Terrain Classification for Outdoor Automobile Navigation — 175

9.1	Introduction	176
9.2	Related Works	178
9.3	2D/3D Coarse Calibration	179
9.4	Feature-based Classification	185
	9.4.1 Depth Image Segmentation	186
	9.4.2 2D/3D Feature Fusion	187
	9.4.2.1 3D Features	188
	9.4.2.2 2D Features	190
9.5	Experiments and Results	192
9.6	Conclusion	195

10 Conclusions — 197

10.1	Summary	197
10.2	Discussion	201
10.3	Direction for Future Work	204

Appendix A - Expert Concerns and Rebuttal — 209

References — 217

CONTENTS

List of Figures

2.1	The experimental platform AMOR	34
2.2	Autonomous mobile robot with LiDAR and CMOS camera mounted near each other to form a 2D/3D coupled system.	36
2.3	Autonomous mobile robot with TYZX DeepSea Camera mounted at the front for 3D scene visualisation.	37
2.4	Autonomous mobile robot with LiDAR, CMOS camera, and MultiCam mounted at the front up.	38
2.5	Triangulation Configuration	38
2.6	Geometric Model	39
2.7	MultiCam	44
2.8	Optical setup of the MultiCam	44
2.9	Examples of MultiCam's images (from left to right): 2d; modulation; depth; infrared intensity. Those images were captured around the campus Hölderlin of Universität Siegen.	47
2.10	Depthmaps (the second row) with the corresponding pictures (the first row), gray values show the depth of the images. Those images were captured around the two campuses Hölderlin and Paul-Bonatz of Universität Siegen.	49
2.11	(a) 2D image. (b) original depth (Best viewed in colours: orange(near); green(neutral); purple(far); white(very far). In the same colour: the darker the nearer). (c) corrected depth by proposed algorithm. (d) 3D scene reconstructed	53

LIST OF FIGURES

2.12 **Left**: An image of a scene. **Middle**: Simple cuts to construct 3D scene from one single 2D image. **Right**: over-segmented image where each small region (superpixel) lies on a plane in the 3D world. ... 54

2.13 A best fit plane for a set of given 3D points. 55

2.14 a) Gray-scale image. b) Raw depth (Best viewed in colour, the colour code is orange: near; green: far; purple: very far, for each colour: the darker the nearer). c) Over-segmented image. d) Mapping regions of interest where the contours of segmented regions are marked in blue colour. 58

2.15 (Left) Raw depth. (Right) Depth refined. 59

2.16 The first row describes 2D images. The second row show the corresponding raw depth data. The last row demonstrates the depth correction given the proposed algorithm (Best viewed in colours: orange(near); green(neutral); purple(far); white(very far). In the same colour: the darker the nearer). 61

2.17 The first row describes colour images where each image consists of red, green and blue channels. The second row shows the corresponding infrared images. 64

3.1 Absorbance Spectra of Chlorophyll a (green) and b (Red) [Asner, 1998] 66

3.2 Reflectance Spectrum of Green Leaf [Asner, 1998]. 67

3.3 Absorption and Reflectance of Green (Left) and Brown (Right) Vegetation [NASA, 2012]. 68

3.4 Scatter plot of NIR reflectance vs. Red reflectance for all pixels in a typical image. Different regions in the scatterplot clearly correspond to different types of pixels in the image. Pixels in the green region correspond to vegetation, and pixels in the blue region correspond to sky [Bradley et al., 2007]. 71

xxiv

LIST OF FIGURES

3.5 Illustration of variations in viewing and illumination conditions for real-world scenes containing vegetation. The vegetation varies in imaging scale and are imaged under different outdoor lighting conditions (Samples of the data can be downloaded here: http://duong-nguyen.webs.com/vegetationdetection.htm). 74

3.6 Examples of our vegetation detection result compared with thresholding NIR and NDVI. 76

3.7 **(Left)** The impact of Luminance on NIR and Red reflectance (normalised grayscale correlation) in vegetation areas. **(Right)** Vegetation samples are sketched on the space NIR-Red as green circles, the impact of Luminance on NIR reflectance is referenced as the blue line. 77

3.8 Vegetation spectra curves in NIR-Red wavelength space as predicted by the adjusted normalized difference vegetation index (in grayscale). The region bounded by the green and Red lines indicates the range of the most popular separated curves used for vegetation detection. . . 78

3.9 Positive relationship between the Modification of Normalized Difference Vegetation Index and the factor A. 78

3.10 The first row illustrates original colour images. The second row shows the results given by NDVI approach. The third row demonstrates the results given by the proposed approach. 80

4.1 The proportion of size of CMOS image to depth image's is equal to the proportion of aperture of CMOS to LMS221's, in each dimension. The 3D model is created by Johannes Leidheiser, Lars Kuhnert and Klaus-Dieter Kuhnert, see more in Leidheiser [2009]. 89

4.2 a) CMOS image, b) cropped CMOS image c)depth image d) segmented image. 89

4.3 Example of reconstructed 3D scenes. 90

4.4 a) an example of vegetation regions extracted from the section III. b)Raw hsv image c) hsv image after thresholding Value's intensities. 95

4.5 Histogram-based retrieval effectiveness for vegetation. 98

LIST OF FIGURES

4.6   Some vegetation detection results obtained from the proposed method.   99

5.1   (a) Optical set-up of the MultiCam. (b) System set-up.   104

5.2   (a) Examples of reconstructed 3D scenes where the exposure-times of 2D and PMD sensors are set at 10 ms. (b) Example of vegetation detection based on thresholding NDVI values where the green colour represents living vegetation, cyan colour denotes dead grass or wet soil. If giving a threshold: $T = \frac{NIR-Red}{NIR+Red} \rightarrow NIR = \frac{1+T}{1-T}Red$, this is a line passing through the origin with the gradient $\frac{1+T}{1-T}$.   107

5.3   Top-left: colour image; Top-right: segmented image; Bottom-left: unstructured points extracted; Bottom-right: texture map is obtained by weighting the average intensity of Gabor responses by the percentage of unstructured points inside the region.   111

5.4   Examples of vegetation detection results obtained from our approach. The first three images are captured with the camera positioned as in **Fig. 1(b)**, when the robot goes (a) down slope, (b) up slope, (c) on flat road. The last image is captured when the camera is positioned horizontally.   114

6.1   From left to right: an original image; near-infrared image; texture image created by the prosed approach; vegetation marked by the proposed algorithm.   119

6.2   The figure shows five examples of multi-spectral data and results. The first column contains original images. The second column shows near-infrared images. The third column illustrates vegetation detection results using the NDVI. The last column demonstrates vegetation detection results using the MNDVI.   136

6.3   The opponent colour space (left) is obtained by rotating the RGB colour space (right) and swapping two channels R and G.   137

6.4   Gabor filter kernels in different scales in rows and orientations in columns.   137

LIST OF FIGURES

6.5 From left to right: original image; segmented image; unstructured texture intensity; confidence map. 138

6.6 Vision-based spreading algorithm. Seed pixels are marked as dark green while the others are white. (For interpretation of the references to colour in this figure legend, the reader is referred to the electronic version of this dissertation.) 138

6.7 From left to right: colour image; NIR image; spectral reflectance-based spreading mask; vision-based spreading mask. 139

6.8 A model of our autonomous mobile outdoor robot. 139

6.9 The first row shows original images. Segmented images are illustrated in the second row. The third row shows the unstructured texture intensities. The fourth row presents the confidence maps. The last row demonstrates the results given by the algorithm. 140

7.1 Example of vegetation detection results given by different vegetation indices. The first column illustrates original images. The second column describes detection results given by the NDVI approach. The third column shows results of MNDVI approach. The last column demonstrates the results from VI_{norm} approach. 149

7.2 The AMOR model is shown here where six blowing devices corresponding with six pipes are mounted at front of the robot (figure provided by J. Schlemper). 150

7.3 The first column describes original image and vegetation detection by by VI_{norm}. The second column shows accumulative background subtraction using Mean & Threshold without and with motion compensation, respectively. The last row illustrates accumulative background subtraction using Mixture of Gaussians without and with motion compensation, respectively. 153

7.4 Block Diagram of the Proposed Algorithm. 156

LIST OF FIGURES

7.5 The first row, from left to right, illustrates original, background subtraction, optical flow and result images, respectively. The second row and fourth row show original images while the third row and the fifth row describe the outputs from our algorithm, respectively. The green and dark green colours reveal passable and non-passable vegetation detected in the result images, respectively. 157

8.1 An example of 3D point cloud given by SICK LMS221 where a) colour image of the scene; b) 3D points in Cartesian coordinate (the maximum distance set is 16 meters, so all farther objects which are not in the case of consideration are illustrated by vertical lines with distance of 18 meters); c) Point cloud triangulation; d) 3D reconstruction of the scene with invalid faces removed. 162

8.2 The first row shows colour images of the viewed scenes. The second row illustrates the corresponding results from point cloud segmentation (best viewed in colours). 166

8.3 The first row shows colour images of the viewed scenes. The second row illustrates the corresponding results from point cloud segmentation (best viewed in colours). 167

8.4 Mapping from 3D point cloud to an array of neighbour pixels. The selection of M (= 4) neighbours pixels in the 3D point cloud is actually taken place by capturing an interval of four numbers in the array, so called prototype point. The new prototype point is one pixel shift of the previous one. 169

8.5 An example of 3D reconstruction of a 3D point cloud delivered by the SICK laser LMS221. The scene consists of flat area, grass, tree and wall. 172

8.6 An example of 3D reconstruction of a 3D point cloud delivered by SICK laser LMS221. The scene consists of building (at right hand), tree and flat area. 172

xxviii

LIST OF FIGURES

8.7	Example of data post-processing for the 3D point cloud in **Fig. 8.5**. The green colour denotes for vegetation areas, the dark blue colour denotes for linear structure areas, and finally the violet colour denotes for solid surface areas	172
8.8	Example of data post-processing for the 3D point cloud in **Fig. 8.6**. The green colour denotes for vegetation areas, the dark blue colour denotes for linear structure areas, and finally the dark cyan colour denotes for solid surface areas	172
9.1	Geometric model of Laser Scanner and CMOS scene planes.	181
9.2	Putting points from LS scene onto the grid plane per line.	182
9.3	a) 3D chessboard model for Laser Scanner and CMOS camera calibration [Leidheiser, 2009]. b) Sketching planes from the centre of the searching window in different levels.	183
9.4	Examples of calibration results.	184
9.5	Examples of segmentation results.	186
9.6	Examples of classification evaluation (in percentage) when applied Multi-classes SVM where seven features are used.	193
9.7	Examples of classification evaluation (in percentage) when applied One-against-all SVM where some specific features are used to detect a particular object. *Note: Road is concrete and we also use elevation information in order to detect roads.*	194
10.1	Monocular setup for the new multi-spectral system.	207
10.2	Rough monocular setup for the new multi-spectral system.	208
10.3	Stereo setup for the new multi-spectral system.	208

LIST OF FIGURES

List of Tables

2.1	Data Sheet of SICK LMS221	42
2.2	Characteristics of Elements	52
2.3	Depth Correction Accuracy	61
2.4	Comparison	61
3.1	Confusion Matrices for Different Methods(%)	80
3.2	Evaluation of Vegetation Detection performances against environmental complexities (EC), illumination complexities (IC), and real-time constraint	81
4.1	Characteristics of Elements	88
4.2	Six extracted features	98
4.3	Confusion Matrix (%)	99
5.1	Confusion Matrices for Different Feature Sets (%)	113
6.1	Confusion Matrices of Different Approaches for Different Groups of Scenes	141
7.1	Confusion Matices of Different Vegetation Indices	148
7.2	Confusion Matix of Passable Vegetation Detection	156
8.1	Classification accuracy	173
9.1	Precision and Times	194

LIST OF TABLES

Chapter 1

Introduction

1.1 Motivation

Autonomous mobile outdoor robots which can drive autonomously in cluttered outdoor environments have received a good deal of attention in recent years. Various agencies of the US Department of Defence have become major sponsors of research in this field through DEMO I, II and III projects; DARPA Grand Challenges. The autonomous off-road robot is foresee able being employed not only in military operations, but also in civilian applications such as wide-area environment monitoring, disaster recovering, search-and-rescue activities, as well as planetary exploration. Different robot systems have been deeply investigated to cope up with a number of challenging problems in domains needed to be solved as different as perception, environment modelling, reasoning and decision-making, control, etc. Possibly the biggest technological challenge for these systems is the ability to sense the environment and to use such perceptual information for control. Indeed, even if equipped with a Global Positioning System (GPS) and an Inertial Measurement Unit (IMU), a robot still needs additionally reliable environment sensing for autonomous operation beyond the line of sight of the operator. Also relying purely on self-localisation could not lead to a safe and reliable autonomous navigation. First, the resolution of GPS or prior environment maps is too low for tasks such as obstacle avoidance. Second, the elevation information given by GPS is not accurate. Third, the infor-

1. INTRODUCTION

mation provided by the prior environment maps is easy to become obsolete. Thus, environment sensing is essential for any autonomous navigation tasks, especially in complex outdoor scenarios. It should be clear that driving in outdoor, non-urban environments has to deal with more complications, such as natural terrain, lighting changes, variety of unknown materials, and other uncertainties, than driving indoors or in urban scenarios. In an indoor environment, one may expect the ground surface in front of the robot to be planar, which helps detect obstacles as something sticking out of the ground plane. In addition, the colour and texture of objects are often persisted from different viewing angles, which leads to ease the object recognition task based on visual features. In contrast, a traditional definition of a lethal obstacle as an object which is rigid and has significant height fails totally to deal with vegetation-like objects, for example tall grass, tuft and small bushes, which are actually passable in real world navigation tasks. This issue is more critical in the case of operating in a corn field, where the robot is not able to move due to all paths blocked by dense geometric obstacles (tall grass, small bushes). Also, on a bumpy dirt road the robot should constantly determine which bumps and holes are small enough to be negotiated and which ones should be avoided. Other challenging situations include the illumination effects such as under/over exposure, shadow, shining, or presence of negative obstacles like ditches, elements such as water, mud or snow, and bad atmospheric conditions such as fog.

Taken all the above into account, this thesis addresses a set of perception tasks that are at the core of any control system for efficient autonomous navigation in outdoor environments. More precisely, we introduce new algorithms for (1) terrain classication and vegetation detection (2).

(1) Terrain classification without doubt is of utmost importance for autonomous navigation especially in the recognition of a traversable or non-traversable terrain, thus attracting numerous studies in robotics. However, existing algorithms apply mostly to urban or indoor environments and do not work well under off-road conditions. This is because typical assumptions about the scene, such as the existence of at ground surface, do not hold in this case. Therefore, this thesis presents new approaches which investigate both 3D spatial distribution and visual features, and fuse

1. Introduction

them to result in a robust terrain classification in outdoor environments. Feature-based approaches proposed instead of the pixel-based enable our algorithms to capture much more detailed object features than does prior art, and thus lead to a more robust classification mechanism.

(2) The current terrain classification techniques mostly consider all obstacles as rigid and static, which, however, fails totally to deal with vegetation-like obstacles. For example, the appearance of a branch of leaves or tall grass looks exactly like lethal obstacles with respect to conventional views, that the vehicle needs to avoid. Such unnecessary avoidance problems coming up for the robot potentially lead to a situation of off-road driving or task-rejection in complex outdoor environments. Therefore, a fully-functional navigation system working outside essentially has to be equipped with a vegetation detection module. Surprisingly, the aim of exploiting the mobility of an autonomous ground vehicle (AGV) has been accelerated very soon while the consideration of affection given by the presence of vegetation in the vehicle's way seemed to be ignored or just few works done. Meanwhile, the presence of vegetation is almost everywhere in the nature as well as its affection on the mobility capability of the autonomous ground vehicle is huge. Locating vegetation areas in a scene helps not only to determine which traversable way to pass but also to understand the local environment for a re-localisation purpose worthy of use in the case of Global Positioning System (GPS) loss. Also, driving on grass or leaves for example would increase wheel slippage, which causes errors in the odometry. Hence, vegetation lets the robot know which types of terrain it is dealing with, and thus which strategies should be applied. In fact, only when the task of forest exploration was given to autonomous ground vehicles recently, a large amount of researches started focusing on vegetation detection, whereby the aim of mobility has switched to the next higher level, from querying road or obstacle to which obstacles can be driven over and which need to be avoided. Nevertheless, to approach a solution for the problem of vegetation detection, a diversity of ways was proposed, underlining different techniques and different models on using different sensors and so on. In order to evaluate those approaches, a structured overview of vegetation detection should be shown beforehand. Therefore, this dissertation reviews the remarkable works done

1.2. PROBLEM DESCRIPTION

for vegetation detection in outdoor navigation in a structured way. Alternatively, due to limitations of the available approaches where a trade-off usually needs to be made between precision and processing-time, a real-time and robust vegetation detection system is still infeasible. Hence, this thesis addresses novel approaches in order to result in a real-time and robust vegetation detection system.

1.2 Problem Description

1.2.1 Terrain Classification

According to the literature of robotic research, terrain classication is generally categorized as vision-based, reaction-based or a combination of vision and reaction-based methods. Vision and reaction-based approaches are analogous to a human driver's recognition of a terrain based on what is seen visually and felt through the vehicle's reactions during traversal of the terrain. The reaction-based method is mainly used for generally classifying different types of terrains, such as soil, mud, concrete, rock, and snow. This is commonly based on estimating the vibration of the robot as well as the resistance from the wheels through traversability. Such classification only helps for speed control and has been done successfully using the available terrain models. The more important and crucial aspect is to visually classify a terrain into traversable or non-traversable parts, or in advance into many different object types such as ground, surface (wall, building), linear structure (wire), positive scattering structure (barbed wire), negative scattering structure (tall grass, small bushes), etc. Consequently, an optimal path to go is computed based on the classified terrain. The later method is called vision-based terrain classication and typically performed using cameras or laser rangefinders. The traditional vision-based terrain classification relies solely on analysing 3D distribution of point clouds given by a LiDAR, or stereo cameras. Meanwhile, scene interpretation based purely on geometric point of view is very difficult, even for human experience and knowledge. Indeed, let's start with very general issues as many objects exist in quite similar 3D structures, so it seems to be impossible to classify them solely by point cloud analysis; when

1.2. Problem Description

two or more objects are near each other, they appear as one in the point cloud; many complex objects like vegetation might exist in different shape, so it is not possible to build common 3D models for them; etc.

Particularly using a LiDAR, one might face more problems due to the scattering effect of beam scattering angle. Whereby, lacking information of far objects usually causes mis-classification. Furthermore, the LiDAR has to sweep up and down to scan the environment, which is extremely time-consuming to acquire the whole frame of point cloud. This hinders many real-time applications.

On the other hand, even though stereoscopic imaging provides both colour and distance information, it also costs much time for calibration and rectification in order to obtain a good depth. The effect of light changes in outdoor is really huge for such the approach, thus its performance behaves differently in different lighting conditions. Especially, this approach can not be used at night time. Overall, the depth produced by the stereoscopic imaging techniques is not that trustable to be used for the safe navigation, and thus needs more investigation and innovation.

Therefore, this dissertation tries to answer the question if it is possible to improve the speed of the LiDAR so that it can be used for real-time applications. Also, we would like to clarify if it is also worth to combine 2D and 3D approaches or fuse colour, texture with 3D distribution information in sense of producing better feature vector components to train object classifiers.

1.2.2 Vegetation Detection

Regarding vegetation detection, it is intuitively trivial for human eye, but not at all for the robot eye. Human eye is able to recognize reflectance changes without considering shadows and unexposed effects; contrariwise, using image processing techniques, an increasing or decreasing in reflectance could happen under different lighting conditions. Indeed, regarding the view-point of image processing, first, there are no specific shape and texture of general vegetation. Second, although vegetation normally owns typical colours such as green, red orange, and yellow, the colour descriptor-based vegetation detection is unstable due to light colour and light inten-

1.2. PROBLEM DESCRIPTION

sity changes under different sunshine conditions in outdoor environments. So, it should be made clear that many publications regarding pattern recognition mentioning grass/leaf detection successfully by using texture and colour information, they however were indicating some very specific species of vegetation but not vegetation in general. As a consequence, those approaches were just applied for robots operating in structured environments but not cluttered ones as investigated in this work. Overall, the only use of colour and texture information cannot result a robust vegetation detection in complex outdoor environments, which drives researchers to come up with the other distinct features rather than colour descriptors, or combine many of them. This might flash an ideal in one's mind back to use laser scanner data, which is very stable and precise against lighting changes. Nonetheless, interpreting point cloud is really challenging. Discriminative features might be extracted from investigating 3D object structures, but currently it is not feasible to use them to robustly detect vegetation. A fusion approach of visual and spatial features might be a good way to go, but a 2D/3D calibration problem needs to be solved beforehand. The complexity of a calibration process between two different vision systems, together with the computational expensive in extracting 2D and 3D features as well as fuse and train them to result in vegetation classifier would make the final solution practically complex or even infeasible for some real time applications.

Alternatively, vegetation is recognised as a visible light absorption specie, especially with red and blue bands. The cell structure of the leaves, on the other hand, strongly reflects near-infrared light. Consequently, the ratio of radiances in the near-infrared (NIR) and red bands has been used as a measure of vegetation index in the satellite remote sensing field. Many different vegetation indices have been derived based on such relationship of spectral reflectance in the two bands. Even though those vegetation indices have been widely and successfully used in many remote sensing applications, for example classifying and positioning the green areas of the earth surface, it is still a problematic thought to apply them directly for mobile robotics applications due to drastically different view-points. Regarding to autonomous ground navigation, there would be more complications to deal with, such as illumination effects (shadow, shining, under-overexposure), views of sky, and presence of variety of different materials, from which the reflected light can have a spectral distribution that is different from that of the sunlight. Indeed, the performance of vegetation indices degrades sharply when an irregular illumination occurs, while illumination effects are inevitable in outdoor environments. Hence, the only use of vegetation indices, or standard multi-spectral approach, is not reliable for vegetation detection in the real world navigation. Finally, since the visible light absorption property of vegetation is indicated by the amount of chlorophyll inside the leaves, one might raise a question if such vegetation indices are really useful in detecting different species/types of vegetation, whose amount of chlorophyll in their leaves diverges considerably. Intuitively, dying vegetation (usually appeared in yellow, brown or red colour) contains very little chlorophyll. This makes the problem of detecting vegetation in general become ever challenging.

1.3 Goald of this Thesis

The purpose of this dissertation is to address the challenging problem of autonomous navigation in cluttered outdoor environments, and to present new ideas and approaches in this newly emerging technical domain. The thesis surveys the state-of-the-art, discusses in detail various related challenging technical aspects, and addresses upcoming technologies in this field. The aim of the thesis is to establish a foundation for a broad class of navigation methodologies for indoor, outdoor, and exploratory missions.

Two main topics located on the cutting edge of the state of the art are addressed, from both the theoretical and technological point of views: terrain classification and vegetation detection.

Terrain classification is studied in a sense of interpreting the surrounding environment as deep as possible. The knowledge about the local environment will be the key factor for robust decision making, especially under uncertainty. Concretely, we goal to classify a terrain into different object types, such as ground, smooth surface (wall, building), rough surface (human, tree trunk), linear structure (wire), positive scattering structure (barbed wire, wired fence), negative scattering structure (tall grass, small bush, canopy). For that aim, different sensor systems and techniques are thoroughly researched.

Alternatively, vegetation is treated as a special object to be deeply studied in this thesis due to a crying need of vegetation detection supporting autonomous navigation in cluttered outdoor environments. Up to now, there is no terrain classification or obstacle avoidance technique, which can work properly under presence of vegetation while vegetation exists everywhere in most of outdoor scenarios. The most reliable way to deal with terrains containing vegetation is to equip a robust vegetation detection module to locate vegetation areas in the scene, and thus provides suitable strategies to cope up with. Nevertheless vegetation is really a complex object so that there still exists no complete solution for such detection task. Therefore, the gold of the thesis is to get the problem solved completely.

1.4 Novel Contributions of the Thesis

The thesis provides five novel contributions with respect to the task of vegetation detection, and two novel contributions for terrain classification. Besides, one contribution of depth correction regarding the data acquisition of Tyzx DeepSea stereo cameras is also shown.

1.4.1 Fitting plane Algorithm-based Depth Correction for Tyzx DeepSea Stereoscopic Imaging

First, it should be clear that the thesis aims to solve two problems: vegetation detection and terrain classification. However, those problems become ever challenging in cluttered outdoor environments, so that many different approaches have been investigated as well as different sensor systems are also used to enable possible solutions. Wherein, a new stereo vision system is presented, the so-called Tyzx Deep stereo camera. Unlike a common stereo camera, the Tyzx Deep stereo camera is

1.4. Novel Contributions of the Thesis

equipped with a hardware module for calibration and rectification processes, thus a ready depth image is achieved from the output of the camera. Due to the aim of fast producing depth images and easy implementation, the calibration algorithm used is the standard block-matching, which is not at that best performance compared with the adjusted block-matching or propagation. In return, the frame rate of the camera is up to 60 fps. With respect to our applications where the robot's maximum speed is at about 3 m/s in an autonomous mode, we do not really need that fast speed of image acquisition but robust depth information. Therefore, the work presented in this contribution deals with the poor performance of depth image generation given by Tyzx stereo vision system under different lighting conditions in both indoor and outdoor environments. For that aim, we introduce a fitting plane algorithm to correct distance information as well as to fulfil the missing points in the original depth. First, the colour image is over-segmented into many small homogeneous regions of interest. Those small regions can be approximately considered as planar surfaces which form the 3D scene. Since 3D points inside each small region should found a plane, this insight is then used to enhance the depth image. Indeed, our algorithm starts with the best fit plane which is built based on the geometric distribution of all 3D points inside each small region, where the sum of all distances from those points to the plane is minimum. When the plane has been built, a 3D point is considered as a defect one if and only if its distance to the plane exceeds three times the average distance of all the points inside the region to the plane. All the defect points will be removed. The new best fit plane will be built with the remaining 3D points, and consider if there still exists defect points. The process is repeated until there is no defect point found. As a result, the last plane is the so-called fitting plane, which is later on used to fulfil all the missing points in the region.

However, the above technique is just applied for textured regions whose depth information is available from the output of the Tyzx DeepSea camera. Due to the fact that there is no depth information on a uniform region from stereoscopic imaging techniques, we presents also in this contribution a new method to overcome this issue. First, the over-segmentation of the colour image results in many small regions where edges are segmented as small regions. In the other words, the neighbours of a

1.4. NOVEL CONTRIBUTIONS OF THE THESIS

uniform region are edge regions. Meanwhile an edge region is textured one, which should have depth information available. Thus, our algorithm starts another loop for depth correction in a uniform region, which relies on the depth information from the neighbours.

Finally assuming that the environment is made up of a number of small planes, we certainly make no explicit assumptions about the structure of the scene; this enables the algorithm to cope up with many different scenes even with significant non-vertical structure. The algorithm has been confirmed to be easily implemented and robust throughout many experiments in different lighting conditions and different scenarios in both indoor and outdoor environments. Concretely, the proposed approach enables a 3D reconstruction capability using Tyzx DeepSea G3 vision system which is infeasible from the raw depth data. Moreover, the proposed algorithm improves more than 48% of 3D reconstruction accuracy compared with the original result given by the stereo vision system over testing 611 scenes under real-time constraint.

1.4.2 Vegetation Indices Applied for Vegetation Detection

Vegetation normally absorbs red and blue light for the photosynthetic process, while it strongly reflects near-infrared light due to the cell structure of the leaves. Hence, vegetation indices are defined as combinations of surface reflectance at two or more wavelengths designed to highlight this particular property of vegetation. There exists many vegetation indices which have been derived to detect vegetation in very different conditions and purposes. So, an overview of available vegetation indices for vegetation detection is shown in this contribution, in order to make clear the advantages and disadvantages from such a multi-spectral approach. At the early state of this work, we tried to exploit some typical properties of vegetation such as homogeneous colour (green, orange, red, or yellow), scatter structure (porous volume) regarding spatial distribution, and distinctive light absorption spectrum (absorb more red and blue band, reflect strongly the NIR band from 800 to 1400 nm). Thus, we used the Tyzx Deep Sea stereo camera where the left eye is covered by a NIR-

1.4. Novel Contributions of the Thesis

blocking filter and the right eye is covered by a NIR-transmitting filter, in order to obtain colour and NIR images, respectively. The NIR and colour images can be used to compute vegetation indices. Nevertheless, both colour and NIR information are not stable in outdoor environments, especially with respect to light intensity and light colour changes. The huge impact from the sunlight degrades the performance of the available vegetation indices-based vegetation detection approaches. Concretely, the changes of NIR and red reflectance are not linear and unpredictable under different sunshine conditions, and thus vegetation indices behave very differently, even for the normalized difference vegetation index (NDVI). Different ways have been suggested to overcome this issue, such as using polarised filters, high dynamic range cameras, etc. Yet, the problem is still unsolved. In this contribution, we would like to introduce a novel and efficient method for vegetation detection against illumination effects by using an independent NIR lighting system. The independent light helps to stabilise the NIR reflectance, then the exposure of the colour sensor can be adjusted as an off-set of that of the near-infrared sensor. This really reduces the impact of lighting changes on vegetation indices. Unfortunately, the use of the additional lighting system affects the relationship between red and NIR reflectance of vegetation, so that traditional vegetation indices can no longer classify robustly vegetation and non-vegetation. For example, NDVI detects only chlorophyll-rich vegetation and all dark materials. Interestingly, a measure on the changes of red and NIR in terms of luminance shows an approximately linear proportion of luminance to red but a logarithm proportion to NIR. On the other hand, NIR-Red wavelength space is sketched with selected vegetation points extracted from 1000 outdoor scenes captured in both morning and afternoon conditions. The distribution of vegetation points is in the top-left part in the space, and a hyperplane to classify vegetation and non-vegetation points is in logarithm form. This confirms a logarithm relationship between the red and NIR information of vegetation against illumination changes. As a result, a modification of normalized difference vegetation index (MNDVI) is derived. The MNDVI has a similar mathematics form as of NDVI, except the red is replaced by log(red) in the formula. The logarithmic term in the later formula expresses the less impact of the red when an artificial lighting system is used. In order to evaluate the

1.4. NOVEL CONTRIBUTIONS OF THE THESIS

performance of MNDVI and other vegetation indices, our autonomous ground vehicle took 5000 raw images and 20 videos of outdoor scenes containing vegetation, under both morning and afternoon conditions as well as shadow, shining and underexposed effects taken into account. Overall, our approach shows out-performance compared with others when taking all environmental and illumination complexities as well as real-time constraint into account. Regarding the performance of the MultiCam, the range measurement is still poor in outdoor environments, thus, the proposed approach could not use depth information for any detection application but just for obstacle avoidance. Alternatively, the wavelength of the modulated light in the MultiCams lighting system strongly focuses on the band around 870 nm while the expected band starts from 800 to 1400 nm, so the chlorophyll less-vegetation like orange/yellow grass is not well detected. However, if extending the spectral width of the modulated light, it degrades the range measurement of the MultiCam. Therefore, a compromise between range measurement and vegetation detection will be considered in our future works. An idea to produce a similar device only for vegetation detection with full band of 800 nm $\rightarrow$ 1400 nm for the desired lighting system will also be taken into account for a further development of the vegetation detection system for outdoor automobile guidance.

1.4.3 2D/3D Feature Fusion for Vegetation Detection

Regarding visual perception, vegetation is recognized through its typical colours like green, yellow, brown, or red-orange. So, it seems to be quite straight forward to investigate colour descriptors in order to detect vegetation. Nevertheless, the colour information is not stable in outdoor environments due to illumination effects, thus methods purely relied on colour features could not provide a robust detection mechanism for safe navigation.

According to the literature of robotics research, interpreting 3D object structures from analysing the point cloud given by a LiDAR is a common way to classify different object types in the viewed scene. In that way, vegetation is detected as a scattering structure object which is different from linear structure (wire), or surface

1.4. Novel Contributions of the Thesis

structure (wall, building, tree trunk). Since the laser data is quite stable and robust, this approach has been applied widely in autonomous navigation. Due to the extreme challenge in geometric distribution-based environmental interpretation, the accuracy of such the approach is not high. For instance, it is impossible to distinguish between dense geometric objects (barbed wire) and vegetation in many cases.

This contribution enables a 2D/3D feature fusion approach which can utilize the complement of three dimensional point distribution and colour descriptor. First, point cloud is segmented into regions of homogeneous distance, which are considered as objects. Local point statistic analysis is then applied to classify the objects into three types of structure: linear, surface, and scatter. Indeed, principle component analysis is applied for all 3D points inside each region, as a result, three eigenvalues and three eigenvectors are computed. The main idea is that: a linear structure object should have only one dominant direction, so the first eigenvalue should be much superior to the others; a surface one should have two dominant directions, thus the first two eigenvalues should be much superior to the last one; a scatter one should have no dominant direction, hence three eigenvalues are not much different.

Second, a 2D/3D calibration needs to be implemented in order to map those segmented regions into the corresponding colour image. Even though, there are many researchers attempted to do full-calibration of coupled vision systems such as Fisheye Laser Scanner and CCD camera or CMOS camera, the result shows mean performance while the cost of computation is very expensive. The precision of reconstructing 3D model drops sharply with the presence of vegetation. One of the main reasons is that interest points are not stable due to the vibration of vegetation. In fact, for the aim of detecting vegetation, we do not need a very precise calibration. A simple 2D/3D mapping with all large objects reconstructed is sufficient. Therefore, we on the other hand propose a simple but fast and efficient 2D/3D mapping technique for the coupled systems: laser scanner and CMOS camera. The technique relied on the insight: if the CMOS camera and laser scanner are positioned near each other in a vertical line, and when objects are far enough, the views from CMOS camera and from laser scanner are nearly the same in a narrow angle. So, a coarse calibration can be done by mapping two images lied on two parallel coordinates. Interestingly,

1.4. NOVEL CONTRIBUTIONS OF THE THESIS

the implementation of this technique is very fast and the robustness is reasonable.

Third, the colours of corresponding objects are obtained by 2D/3D mapping after the coarse calibration. In fact, the colour invariant descriptors have been evaluated individually in the precious art, where they show high precision of detecting specific objects such as aeroplane, person, horse, and car. However, the detection of vegetation, in particular potted plant, is still very poor, at about 0.2 in average precision. One of the major problems for that is the shift and changes of intensity and colour in different light conditions while the vegetation tends to be recognized based mostly on its colour. So, two of interesting features should be taken into account are the mean and standard variation values of intensity and colour which imply the light condition of the view scene. The interesting point for vegetation images is that the main colour should be theoretically green in the HSV colour space under most different environment conditions. In reality, this is not always true for scenes containing sky. The affection of sky tends to turn the colour of image to red, brown, etc. The issue is often caused by the low intensity of the Value (in the HSV colour space), and thus can be solved by removing the pixels which have too low the Value' intensities. The green or orange colour appears as a majority colour in vegetation images in the HSV colour space. This drives us to come up with a vegetation detection based on colour histogram distribution which is well-known in image retrieval and in detecting homogeneous colour objects. Hence, colour histogram distances including histogram euclidean, histogram intersection and histogram quadratic distances are applied to extract colour features.

Finally, brightness (mean,standard deviation of intensity), spatial features and colour features are fused and then trained by Support Vector Machine (SVM) to generate vegetation classifier. In order to evaluate the performance of the proposed method. 500 different scenes of cluttered outdoor environments are captured by the SICK laser scanner LMS221 with 41x157 pixels resolution and the Logitech QuickCam Pro 9000 with 640x480 pixels resolution, in both morning and afternoon conditions. The maximum distance set is 16 m. 300 pairs of point clouds and CMOS images are used for training and the other 200 are used for testing. The accuracy of the method is 82.86% while the total processing time is at around 2580 ms. It is clear that the approach is not really reliable for on-board navigation. Thus, the main use is to predict the scene categories at the front of the vehicle and interpret the current environment by localising vegetation areas around. In reality, outdoor autonomous navigation has to face with unknown environments and unknown situations. Whenever, the autonomous robot gets into a tough situation where it could not find which way to go or all paths seem to be blocked by lethal obstacles. In such situation, the approach initiates a solution. The robust detection of the proposed method enables a more interaction between autonomous robots and natural environments.

1.4.4 General Vegetation Detection Using an Integrated Vision System

Although the 2D/3D feature fusion approach provides high accuracy in detecting vegetation, its applicability in autonomous navigation is limited due to its computational expensiveness. On the other hand, the multi-spectral approach can provide very fast detection results by thresholding vegetation indices, but performs differently in different lighting conditions. The fact is that those limitations of the two approaches come mainly from hardware issues. Therefore, an improvement in hardware has to be taken into account. Indeed, we have developed a new platform of SICK laser scanner LMS221 by mounting a light weight mirror directly in front of the 2D laser scanner. In that way, the laser scanner is fixed while the mirror is rotated by a motor to reflect the laser beam for capturing the environment. Hence, a higher velocity can be achieved, which is proportional to the velocity of the motor. Even though, the achievement of 6 Hz frame rate is reached, this is still not fast enough for real-time applications. So, when the time issue is seriously taken into account, the use of the laser scanner is no longer suitable (exceptionally Velodyne is very fast but too much expensive, and thus is not considered in our case). Besides, the stereoscopic imaging is not a solution while it also takes time to result in a depth image. Even worse, the depth information is not precise enough to be used for a statistic analysis process. Therefore, a research on what information is really needed and which vision devices can be worthy used to acquire the information has been done.

1.4. NOVEL CONTRIBUTIONS OF THE THESIS

The spectral reflectance property, without doubt, is the most discriminative feature which classifies vegetation with others, and thus should not be negligible. Although, the light spectral distribution of the sunlight changes sharply due to illumination effects, which might be compensated by using an independent lighting source. Second, regarding visual perception, colour information is the most important element which helps human eye simply recognising vegetation, and thus is deserved to be investigated. Taken all the above into account, this contribution introduces the use of a new vision system integrated from Photonic Mixer Device (PMD) equipped with a NIR lighting system and CMOS camera, the so-called MultiCam. The MultiCam can provide simultaneously near-infrared (NIR), colour, and depth images. Whereby, the reflectance of the modulated NIR given by the PMD sensor and the red channel of the CMOS sensor are used to calculate Normalized Difference Vegetation Index (NDVI). The exposure times of the PMD sensor and CMOS sensor are programmably adjustable, thus the NIR reflectance can be stabilised. A more stable multi-spectral system is achieved when the exposure time of the CMOS sensor is set as an off-set of that of the PMD sensor. Practical experiments reveal that thresholding NDVI could not provide robust vegetation detection, but fusing NDVI, NIR, and brightness gives rather good performance. This comes out the idea of using a classification-based method which might get more advantages when visual features are added. In this contribution, we will derive a methodology for generating colour histogram models and assessing unstructured texture orientation to create visual features.

Alternatively, the MultiCam modulates the NIR light (wavelength centred at 870 nm) to estimate the time-of-flight, and the distance is computed from the phase-shift between the reflected light and the emitted light. This is strongly affected by the sunlight which has a wide spectral range. Particularly, the reflected NIR light from the poor NIR reflectance surfaces and the very far objects has a very low intensity that is confused with the NIR light of the sunlight, and thus might not be reconstructed correctly. Consequently, the phase-shift is not accurately calculated or the computed distance is wrong. Interestingly, vegetation reflects dramatically the NIR light, as a result its depth contains lesser noise than non-vegetation objects. This finding can be

1.4. Novel Contributions of the Thesis

worthy used to create good spatial features for the vegetation classifier. For that aim, a new system setup where the MultiCam is positioned as looking down to restrict the distance of the NIR light travel, and a relative distance estimation method referencing a perfect flat ground is described to obtain quickly 3D point cloud in the vehicle frame, thus, enables a 3D distribution analysis for spatial feature extraction. Finally, NDVI, spatial features and visual features are gathered to form feature vectors, which are then used to train a robust vegetation classifier. In all real world experiments we carried out, with 1000 scenes captured in both morning and afternoon conditions, our approach yields a detection accuracy of over 90%.

1.4.5 Spreading Algorithm for Efficient Vegetation Detection

A classification-based method is presented in 1.4.4 where visual and spatial features are extracted and trained to generate the vegetation classifier. As the general rule, the more features are used the better classifier is achieved. The key limitations of such the classification-based method are the dependence on the dataset; many features need to be extracted and trained to obtain a good classifier, which increases the complexity of the method, and thus restricts its applicability in many real-time applications. Interestingly, visual features are recognised as playing an important role in producing and strengthening the vegetation classifier through cross-validation in the training process. This motivates us to pay more attention on studying colour and texture as well as spectral reflectance property of vegetation. A deep investigation on colour and texture of vegetation has been carried out. Remarkably, there are two interesting findings as follow.

- Although different species of vegetation can have different colours, considering a small region of it, the colour is expected to be homogeneous.

- The textures of most of vegetations are unstructured or turbulent. That can be inferred as we would find many pixels in a small vegetation region, which have different texture orientations with the texture orientation of the region.

1.4. NOVEL CONTRIBUTIONS OF THE THESIS

So if we know a vegetation pixel, we can try to find the connected ones by measuring the colour and texture dissimilarities between the pixel and its neighbours. However such colour dissimilarity measure is variant in the RGB colour space. Thus, an opponent colour space is conducted to achieve the intensity-shift invariance. On the other hand, as stated in the second finding, the unstructured texture is evaluated by firstly estimating the texture orientation of a small region and of all pixels inside the region. The percentage of pixels which have different texture orientations with that of the small region expresses the degree of turbulence of the region. Object which has the similar colour with a given vegetation and high degree of texture turbulence should be judged as vegetation.

Alternatively, a multi-spectral based method is described in 1.4.2 where vegetation indices are applied to detect vegetation. In order to reduce the impact of illumination effects on the performance of the system, an active NIR lighting system is used, and thus a modification of normalised difference vegetation index (MNDVI) is derived. Consequently, the detection mechanism works well and fast in detecting chlorophyll-rich vegetation (which reflects strongly the NIR light and absorbs significantly the red light), in different lighting conditions. Nevertheless, there is a confusion between chlorophyll-less vegetation and warm objects or NIR reflectance surfaces.

Taken all the above into account, this paper addresses a solution for efficient vegetation detection by using a spreading algorithm. We aim to create an adaptive learning algorithm which performs a quantitatively accurate detection, as well as fast enough for a real-time application. Indeed, chlorophyll-rich vegetation can be detected using the multi-spectral approach, and then considered as seeds of a "spread vegetation". Chlorophyll-less vegetation is detected by spreading the spread vegetation based on colour dissimilarity and the degree of texture turbulence. Overall, there are two criteria on colour and texture that we have to deal with. To many experts in the field of machine learning, it seems to be worthwhile to investigate a probabilistic combination of different classifiers in this case. Actually, in an early approach, we already tried to use a Markov Random Field (MRF) to model the visual difference (colour and texture). However, the trained MRF only helps to detect vegetation

1.4. Novel Contributions of the Thesis

which has simultaneously high probabilities of both colour and texture similarities, $MRF = P_{texture} P_{color}$. Thus, the algorithm could not detect vegetation in a dark region where there seems to be no texture detected, or $P_{texture} = 0$; Two vegetation neighbour pixels could not be joined if their colours are to too much different, or $P_{color} = 0$. This hinders the purpose of detecting a variety of vegetation appeared in many different colours. The multiplication in MRF degrades the performance of the algorithm in the case one feature missed. This leads us to the idea of using a convex combination. The convex combination, trained via semi-supervised learning, models the difference of vegetation pixels and between a vegetation pixel with a non-vegetation pixel, thus, allows a greedy decision-making to spread the spreading vegetation, so called vision-based spreading. Hence, the convex combination helps to vote for the candidate which dominates texture similarity or colour similarity, or both of them. Certainly, we acknowledge that based on such convex combination the algorithm is rather greedy. To avoid the over-spreading, especially in case of noise, a spreading scale is set. On the other hand, another vegetation spreading based on spectral reflectance is carried out in parallel. Concretely, we decrease the thresholds of vegetation indices step by step to restrict the possible vegetation areas at which vision-based vegetation spreading can occupy. Finally, the intersection part resulted by both vision-based and spectral reflectance-based spreading is added to the root. The approach takes into account both vision and chlorophyll light absorption properties. This enables the algorithm to capture much more detailed vegetation features than does prior art, and also give a much richer experience in the interpretation of vegetation representation, even for scenes with significant over- or under-exposure as well as presence of shadow and sunshine. Remarkably, the method of pointing out turbulent texture described in this work leads to distinguish between an dense edge region (barbed wire) and an unstructured texture region (vegetation), which is infeasible in previous works. Consequently, the proposed method outperforms others; this is pointed out in a concrete evaluation on the performances of all vegetation detection approaches in this contribution.

1.4. NOVEL CONTRIBUTIONS OF THE THESIS

1.4.6 A Novel Approach for a Double-Check of Passable Vegetation Detection in Autonomous Ground Vehicles

Even if we could detect robustly vegetation, there is still a concern whether the vegetation is passable or not, especially in the case the vegetation is at the front of the robot. A good passable vegetation detection enables a safe autonomous navigation in cluttered outdoor environments. While many publications in the remote sensing field have reported that the more chlorophyll vegetation has the easier it is to drive through, it is still problematic of how to estimate accurately the amount of chlorophyll inside vegetation. Approximately, vegetation indices indicate a relative amount of chlorophyll existed in the vegetation but not accurate enough to be used in our purpose. Regarding the kinematic consideration, some types of vegetation such as grass and cornstalks are easy to drive through because of less resistance. Indeed grass and cornstalks are soft and movable, which can be clearly seen that they are easy to be moved under blowing wind. Therefore we propose a novel approach for a double-check of passable vegetation detection, which is based on the high amount of chlorophyll in and the less-resistance of vegetation. The approach contains two phases: firstly vegetation in general is detected; secondly blowing devices are used to create strong wind to effect vegetation. Based on motion estimation, detected vegetation being soft or movable is then judged as passable one.

At the first phase, one might use the spreading algorithm which is currently at the state of the art in vegetation detection techniques and is introduced in the previous section. Alternatively, while the aim is to detect passable vegetation at the front of the robot for the purpose of navigation, it is preferable to position the MultiCam looking down to setup the region of interest right at the front of the robot. Within that configuration, the impact of illumination effects is significantly reduced, and thus practical experiments have shown that a convex combination between vegetation indices can also result in a robust and fast vegetation detection mechanism. Indeed, in this particular work, we prefer to use the later method which is much faster than the first one.

At the second phase, moving vegetation after an initiation of blowing devices

1.4. Novel Contributions of the Thesis

has to be detected. There are two scenarios that we should make clear: first, moving vegetation is detected in case of a running robot; second, moving vegetation is detected in a halt state of a robot. At the current state, it is still infeasible to detect moving vegetation in case of a running robot because the vibration from driving on natural terrains and from the robot itself is somehow even bigger than the movement of vegetation. Therefore, we state clearly that this work just aims to detect passable vegetation in a halt state of the robot. That means the robot needs to stop, and then starts the blowing devices. Even in a halt state, the robot itself still has a vibration caused by the engine when operating. Thus, a motion compensation and motion detection techniques have to be applied to detect foreground objects, which are then mapped to the detected vegetation in the first phase to obtain moving vegetation. Furthermore, it is intuitive that the more movement vegetation does the less resistance vegetation own, thus the easier vegetation is to drive through. Hence, this paper also proposes optical flow techniques to record the movement of vegetation.

Regarding to the system design, we need a blowing device to create wind to effect vegetation. One might immediately think about utilising the available air compressor of the robot's air-brake system. This, however, is not a reliable solution. The robot lasts its battery quickly because of high power consumption for the charging process of the air compressor. The blowing duration is very short due to the small air compressor tank. More seriously, using the air compressor would affect to the break system, thus, potentially causes an unexpected movement of the robot. Then, we come up with an idea of using independent blowing devices. Take a look at current products for such work, we find Bosch leaf blowers such as Bosch ALB 18 LI Cordless Li-Ion and Bosch ALS 25 which are really suited for the work and quite cheap, at around 80 Euro. Indeed, the leaf blowers can run continuously for 10 minutes at blow speeds of up to 215 km/h. Meanwhile, the robot only needs to turn on the blowing device in case of facing vegetation as obstacle, and for each time the blowing duration required is just from five to ten seconds. Therefore, after each fully charge, the device can be used for at least 60 halt states, which is so far satisfy us at the current stage.

Finally, in all real world experiments we carried out, with 1000 input images cap-

1.4. NOVEL CONTRIBUTIONS OF THE THESIS

tured from 50 halt states of the vehicle (20 frames per each halt state), our approach yields a detection accuracy of over 98%. We furthermore illustrate how the active way can improve the autonomous navigation capabilities of autonomous outdoor robot.

1.4.7 Terrain Classification Based on Structure for Autonomous Navigation in Complex Environments

Terrain classification is very important in regard to the efficiency and safety of a robot in autonomous navigation. A robot able to classify the terrain ahead can optimize its speed for the terrain (drive slower on a rough terrain and faster on a smooth terrain) or avoid potentially hazardous areas, such as stretches of mud or sand in which a ground-based robot could become stuck. Additionally, one of the goals of this study is also to interpret in details the terrain structure in sense that helps to determine different object types inside the terrain such as flat ground, smooth surface obstacle (wall, building) linear structure objects (wire, branch of tree), scatter structure (tree canopy, needle tree, bushes). This is worthy for decision-making frame-work in navigation. With regard to the literature of robotic research, up to now, the most reliable way to classify terrains is based on analysing the point cloud given by a LiDAR. While most of recent methods for LiDAR processing are purely found on the local point density and spatial distribution of the 3D point cloud directly. Commonly, the 3D-space occupied by the 3D point cloud is divided equally into many voxels (for example, a cube). The number of 3D points inside each voxel reveals the structure of it: surface (many points); empty (no point); scatter (mixture). However, many hand-tuned parameters need to be fulfilled, and thus makes the approach unstable. Indeed, it is still problematic of how many 3D points inside a voxel should be considered as "many" so that the voxel would be judged as having surface structure, while a voxe with different selected sizes will behave differently, and thus requires different hand-tuned parameters. Furthermore, there exists scattering effect of beam scattering angles, whereby the farther an object is from the LiDAR the lesser the number of 3D points obtained about the object is. Alternatively, analysing the 3D

1.4. Novel Contributions of the Thesis

spatial distribution under pixel level has also got a good deal of attention recently. This, however, could not bring much objective information and also failed to deal with complex terrains.

From our perspective, there should be more objective information to view a scene under the object level. Thereby, a segmentation of the point cloud should be done beforehand to result objects in form of regions of interest. For that aim, we propose to apply the Graph-Cut technique which leads to segment the point cloud with respect to homogeneous distances in local regions. Consequently, objects or segmented regions of interest can be classified into different 3D structures through a local point statistic analysis. More precisely, a principle component analysis is applied for a set of 3D points of each object to classify the object into three types of 3D structure: linear (wire, small branches of tree), surface (wall, solid obstacles), scatter (tree canopy, needle tree, bushes). Indeed, regarding geometric distribution, the three returned eigenvalues from the principle component analysis are consistent with the 3D structure of those analysed 3D points: a linear structure indicates only one dominant direction, so the first eigenvalue should be superior to the others; a surface structure implies two dominant direction, thus the first two eigenvalues should be similar to each other and superior to the last one; scatter structure reveals no dominant direction, hence three eigenvalues are similar to each others.

Besides, it is intuitively distinguishable between linear structure and surface objects. Still, there exists some confusion between rough surface and scatter structure objects. Therefore, this section proposes a distance variation estimation to ease the problem. Whereby, the variation of distances of 3D points of a surface object, even a rough surface one, is expected to be lesser than of a scatter object. Due to the imperfect segmentation of point cloud in an early step, it is recommended to eliminate the edge points out of interest for such estimation, thus we would like to call the outcome as distance variation inside edgeless regions (DViER). Through extensive experiments, we demonstrate that this feature has properties complementary to the conditional local point statistics features, and thus together show significant improvement in classification performance.

Regarding to experiments and results, 300 different scenes of cluttered outdoor en-

1.4. NOVEL CONTRIBUTIONS OF THE THESIS

vironments are captured by the SICK laser scanner LMS221 with 81x330 = 26730 pixels resolution and the maximum distance set is 16 m. The angular separation between laser beams is $\frac{1}{4}$ degree over a 90^0 field of view. The angular separation between laser sweeps is $\frac{2}{3}$ of a degree over 120^0. The classification results are evaluated by comparing the output of the classifier with the hand-labeled data. In this paper, we evaluate the discrimination between scatter, linear, and surface structures rather than the specific classes of classification such as grass, trees, bushes, building, roads etc. Actually, if we can have a good classification of the three structures, the object classification can lately be realized by evaluating the relationship between the object structure and the three structures. For example, the grass should be a vegetation area with little presence of linear structure, while the bushes and trees should be vegetation areas with dense presence of linear structure. The discrimination between trees and bushes can be clarified by estimating the elevation of their centroids. The roads and lethal obstacles can also be classified by their elevation regarding the discrimination of solid surface areas. Overall, the proposed approach provides an improvement of around 10% to 17 % in accuracy compared with previous ones.

1.4.8 A Novel Approach of Terrain Classification for Outdoor Automobile Navigation

The aim of this research is to achieve a terrain classification mechanism which is not only to classify a terrain into traversable or non-traversable regions but also in advances to be able to classify many object types inside the terrain, and thus brings much benefit for the safety and efficiency of an autonomous robot operating in cluttered outdoor environments. The fact is that the non-urban environments usually bring much more complications than the urban ones, hence a method relied purely on geometric distribution analysis could not lead to a robust mechanism in many circumstances. Indeed, interpreting a complex scene is somehow infeasible even with human knowledge and experience by only observing the point cloud of it. The main issue is a lack of information. It is intuitively clear that the classification problem should be easier to be solved by having more colour and texture information. There-

1.4. Novel Contributions of the Thesis

fore, a coarse 2D/3D calibration is presented in this paper to connect colour, texture and 3D distribution information of the viewed scene; consequently, a complete 3D scene is reconstructed. The coarse calibration is proposed instead of the full one, because with current computer vision techniques the full 2D/3D calibration does not provide a significant improvement in the accuracy compared with the proposed coarse calibration while requiring a heavy computation, especially for cluttered outdoor scenes.

Regarding to visual perception, most of discriminative features of an object could not be found under pixel level even with all available 2D/3D information. Hence, we suggest to segment the viewed scene into object types or regions of interest. To do so, one might use common segmentation techniques on the colour image, this however does not really provide good results in many circumstances due to the unstable colour information under different lighting conditions. The appearance of shadow is inevitable in outdoor environments In our case, we apply a Graph-Cut technique to segment the depth given by the LiDAR, thus having a more stable result. Indeed, the distances given Laser Scanner are very precise whereby we can obtain a very fine depth image which is quite stable even under complex conditions and environments. Furthermore, the segmentation technique considers both local and global properties of the scene, and thus the results are not either too coarse or too fine.

Overall, 2D features (extracted from colour, texture) and 3D features (extracted from 3D structures, 3D distribution) gathered to form a feature vector. In this work, we have tested different sets of 2D/3D features (for example: RGB-shift, HSV-shift, RGB histogram distances, HSV histogram distances, opponent-shift, geometric similarity, local point statistic features) as inputs for a training process using support vector machine (SVM), in order to point out the optimal vector components for different object classifiers. We aim to classify terrain into six object types: tree trunk, human, wall/building, sky, vegetation, road. There are more objects need to be detected, so we use the multi-classes SVM and one-against-all SVM instead of the binary SVM which is commonly introduced in the literature of object detection-based classification. Consequently, the classifier generated by the proposed method

1. INTRODUCTION

provides 54.3% for detecting tree trunk, 72.5% for detecting human, 58.6% for detecting wall/building, 88.2% for detecting sky, 86.4% for detecting vegetation and 87.5% for detecting road. The result is very impressive and show a significant improvement to the art in terrain classification.

1.5 Document Structure

The remainder of this thesis is organized as follows:

Chapter 2 provides some fundamentals related to background knowledge and hardware systems for a better understanding of the works done in this thesis. Particularly, our autonomous robot as well as different devices and sensor systems used are briefly presented. Since the first main contribution is about improving the data acquisition in stereoscopic imaging, we prefer to put it in the corresponding section in this chapter.

Chapter 3 surveys multi-spectral approaches applied for vegetation detection, whereby proposing the idea of using an active NIR lighting system to achieve a more stable multi-spectral system; accordingly devising the Modification of Normalized Difference Vegetation Index (MNDVI) obtained from a regression analysis of the red and NIR reflectance of vegetation in different lighting conditions. Practical experiments have confirmed that the new modification form is worthy used whenever an independent light source is equipped.

Chapter [4&5] cover vegetation detection based on classification. Concretely by combining data given by both a CMOS camera and a LiDAR, a 2D/3D feature fusion is described to exploit discriminative visual features to robustly detect vegetation (see chapter 4). Secondly, Chapter 5 studies an integrated vision system, which is built from mounting both CMOS sensor and PMD sensor into a molecular setup; and additionally equipped with an active near-infrared lighting system, to acquire simultaneously colour image, near-infrared intensity and distance information. Within this section, those outcomes from the vision system are well analysed to extract the optimal vector components to be trained via support vector machine, in order to robustly detect general vegetation.

Chapter 6 presents the work as a follow-up of the system setup in the Chapter 3. Whereby, a spreading algorithm is devised to efficiently and robustly detect variety of vegetation in different lighting condition as well as under different illumination effects. This approach significantly outperforms the state of the art of general vegetation detection.

Chapter 7 deals with the problem of detecting passable vegetation for navigation in cluttered outdoor environments. The approach relies on the spectral reflectance and the less-resistance properties of vegetation, whereby proposing an efficient model and systematic design for the passable vegetation detection system.

Chapter 8 studies terrain classification based on geometric structure of object types. This is done by applying local point statistic analysis and estimating distance variation inside edgeless regions.

Chapter 9 introduces a novel approach for terrain classification. The approach utilises the fusion of colour, texture and 3D distribution information to deeply interpret the world representation of different object types.

Chapter 10 concludes with a summary, discussion of applications, and thoughts about future directions of research.

Appendix A, named as Expert Concerns and Rebuttal, mentions several interesting questions from Professors/Reviewers/Editors, and our rebuttal for that correspondingly. This part might help readers to deeply get into technical discussion related to this work.

1.6 Publications

The research comprising this thesis has been presented at a number of international conferences and workshops as well as published in international journals. The relevant publications are listed below grouped by the publication types.

- **Journal Articles**

1. INTRODUCTION

1. D.-V. Nguyen, L. Kuhnert, and K.-D. Kuhnert. General Vegetation Detection Using An Integrated Vision system. *International Journal of Robotics and Automation*, ACTA press, to appear 2013.

2. D.-V. Nguyen, L. Kuhnert, and K.-D. Kuhnert. Spreading Algorithm for Efficient Vegetation Detection. *Robotics and Autonomous Systems*, Vol. 6, No. 12, pp. 1498-1507, December-2012.

3. D.-V. Nguyen, L. Kuhnert, and K.-D. Kuhnert. Structure Overview of Vegetation Detection. A Novel Approach for Efficient Vegetation Detection using An Active Lighting System. *Robotics and Autonomous Systems*, Vol. 60, No. 4, pp. 498-508, April-2012.

4. D.-V. Nguyen, Thuong-le. Super-resolution Method Combining Transforms between Frequency Domain and Wavelet Domain. *Journal on Information and Telecommunication*, 5(1):40-48, April-2009.

1. Introduction

- **Conference Papers**

    1. D.-V. Nguyen, L. Kuhnert, S. Thamke, J. Schlemper, and K.-D. Kuhnert. An Active Approach for A Double-Check of Passable Vegetation Detection in Autonomous Ground Vehicles. *The 15th IEEE Intelligent Transportation Systems Conference*, Alaska, USA, Sept-2012.

    2. T. Jiang, D.-V. Nguyen and K. -D. Kuhnert. Auto White Balance Using the Coincidence of Chromaticity Histograms. *The 8th International Conference on Signal Image Technology and Internet System (SITIS2012)*, Naples, Italy, Nov-2012.

    3. T. Jiang, Duong Nguyen and K.-D. Kuhnert. A Flexible Auto White Balance Based on Histogram Overlap. *11th Asian Conference on Computer Vision(ACCV2012) Workshop on Computational Photography and Low-Level Vision*, Korea, Nov-2012.

    4. D.-V. Nguyen, T. Jiang, L. Kuhnert, and K.-D. Kuhnert. Fitting Plane Algorithm-based Depth Correction for Tyzx DeepSea Stereoscopic Imaging. In *International Conference on Communications and Electronics(ICCE)*, Hue, Vietnam, Aug-2012.

    5. L. Kuhnert, S. Thamke, M. Ax, D.-V. Nguyen, K.-D. Kuhnert. Cooperation in heterogeneous groups of autonomous robots. IEEE International Conference on Mechatronics and Automation (ICMA), Chengdu, China, 3-5 August 2012.

    6. L. Kuhnert, D.-V. Nguyen, S. Thamke, and K.-D. Kuhnert. Autonomous explorative outdoor path planning. In *IASTED International Conference on Robotics (Robo 2011)*, Pittsbugh, USA, November 2011.

    7. D.-V. Nguyen, Lars Kuhnert, and K.-D. Kuhnert. An Integrated Vision System for Vegetation Detection in Autonomous Ground Vehicles. In *IASTED International Conference on Robotics (Robo 2011)*,Pittsbugh, USA, Nov-2011.

1. INTRODUCTION

8. D.-V. Nguyen, L. Kuhnert, T. Jiang and K.-D. Kuhnert, A Novel Approach of Terrain Classification for Outdoor Automobile Navigation. In *IEEE International Conference on Computer Science and Automation Engineering (CSAE)*, Shanghai, China, June-2011.

9. Tao Jiang, K. D. Kuhnert, D.-V. Nguyen and L. Kuhnert. Multiple template auto exposure control based on luminance histogram for on-board camera. In *IEEE International Conference on Computer Science and Automation Engineering(CSAE)*, Shanghai, China, June, 2011.

10. D.-V. Nguyen, L.Kuhnert, T.Jiang,S.Thamke and K.-D. Kuhnert. Vegetation Detection for Outdoor Automobile Guidance. In *IEEE ICIT-2011 International Conference on Industrial Technology*, Auburn, Alabama, USA, March-20.

11. D.-V. Nguyen, L. Kuhnert, J. Schlemper, and K.-D. Kuhnert. Terrain classification based on structure for autonomous navigation in complex environments. In *Int. Conf. on Communication and Electronics (ICCE)*, Vietnam, August 2010.

12. D.-V. Nguyen, L. Kuhnert, M. Ax, and K.-D. Kuhnert. Combining distance and modulation information for detecting pedestrians in outdoor environment using a pmd camera. In *The 11th IASTED Int. Conf. on Computer Graphics and Imaging (CGIM 2010)*, Innsbruck, Austria, February 17-19, 2010.

13. M. Langer, L. Kuhnert, M. Ax, D.-V. Nguyen, and K.-D. Kuhnert. 3D object recognition and localization employing an analysis by synthesis system. In *IADIS Int. Conf. on Applied Computing*, pages 132-140, Rome, Italy, November 19-21, 2009.

14. L. Kuhnert, M. Ax, M. Langer, D.-V. Nguyen, and K.-D. Kuhnert. Absolute high-precision localisation of an unmanned ground vehicle by using real-time aerial video imagery for geo-referenced orthophoto registration. In *Fachgesprche Autonome Mobile Systeme (AMS)*, Karlsruhe, Germany, 2009.

15. D. V. Nguyen, Thuong Le-Tien, Sung Young Lee. Constructing computational methods for artifiical plant growing light .In *Int.Conf. on Communication and Electronics*, 56:438–446, Vietnam, June-2008.

1. INTRODUCTION

Chapter 2

Fundamentals

As outlined in the introduction, the investigation on vegetation detection and terrain classification aims for guiding an autonomous robot to fully exploit its mobility capability in both urban and non-urban environments. Therefore, the experimental wheel-based vehicle used for all experiments to test the proposed algorithms for both vegetation detection and terrain classification is briefly introduced in this chapter, please see section 2.1. As mentioned in section 1.2 that the two perception tasks (vegetation detection and terrain classification) are very challenging in outdoor environments, which could not be done efficiently within using a single sensor system. Multiple approaches using multi-sensor systems have been proposed, and thus the principles of the devices or sensor systems are presented, afterwards in this Chapter, for better understanding the thesis content. Concretely, since the thesis goals to interpret not only the 2D (.e.g. colour, texture) and 3D (.e.g. shape, structure) object appearance, but also the object reflectance property in different spectral illuminant, three types of sensor systems are deeply studied, including 2D imaging sensors, 3D range sensors, and multi-spectral sensors. First, the introduction of common 3D range measurement techniques and corresponding devices applied for distance measurement is presented, including Light Detection and Ranging (.e.g. laser scanner) in section 2.2, Structured Light (.e.g. Kinect sensor) in section 2.3, Stereoscopic Imaging (.e.g. stereo cameras) in section 2.5. Section 2.6 describes the principle of multi-spectral imaging systems. While multi-spectral sensor systems are very expensive,

2. FUNDAMENTALS

the section also points out how to build a cheap system which is still worthy used for the purpose of detecting vegetation. Because the 2D imaging sensors (CCD/CMOS) are very popular, which are used in common digital cameras and whose principle are mentioned in any computer vision books, they will not be repeated in this chapter.

2.1 The Experimental Platform AMOR

In order to validate the developed terrain classification and vegetation detection methodologies, a mobile robot capable of traversing rugged terrain is required. The robot should have high power enough to provide the required maneuverability across uneven ground, and thus needs to be large and heavy. However, with the aim to operate autonomously in cluttered outdoor environments, the robot must be small enough to drive through light forests. **Fig. 2.1** shows the experimental wheel-based vehicle AMOR (short for Autonomous Mobile Outdoor Robot), which has been developed at the Institute for Real-Time Learning Systems of the University of Siegen. The mechanical platform of AMOR is a quad-type Yamaha Kodiak 400, a model of All Terrain Vehicle, which offers a superior cross-country performance while having relatively small size. The robot is equipped with various sensors and actuators.

Typical operation terrain: Rugged off-road   Typical operation terrain: Forest

Figure 2.1: The experimental platform AMOR

Steering, throttle, brake and gears can be remotely operated or controlled by

2. Fundamentals

the computer. To capture its state and its surroundings, AMOR is equipped with the following sensors/devices: Laser scanner SICK LMS221 (two front, one rear); Ultrasonic sensors (five forwards); Imaging cameras (one forward-looking, one with centralized visibility); Smart camera; MultiCam; Stereo-camera system; DeepSea Stereo Camera; Inertial sensor systems; Accelerometer; Rate sensor; Optical motion sensor (two units at the rear of the vehicle); Electronic compass; Differential GPS; Weather station. Besides, in some applications, typically in planetary exploration or rescue mission, a microdrone (Unmanned Aerial Vehicle) is mounted on the top of the robot for ground and air cooperation. The size of the robot is $1.2m \times 2.6m \times 1.8m$, and the weight is about $650kg$.

The robot has a powerful computer on x86-based, which is connected to those sensors and actuators via a control area network (CAN) bus. The ATMEGA-32 8-bit micro controller from Atmel has been chosen mainly because of the large amount of free software available for this type (i.e. the free IDE supplied by Atmel supports the free GNU GCC compiler). A key element of the architecture is the VSAL (Virtual Sensor Actor Layer) which unifies the access to all sensors and actors, creates virtual sensors out of one or more physical ones, and allows easy reconfiguration of the system. A specialized sensor configuration description language supports this part. The higher levels posses a hybrid architecture, the so called CAPTAIN (Control Architecture Providing Task Arbitration and Sequencing in Intelligent Robot Navigation), which is formed by a set of functional units and a mission control. Data and event from the functional units will feed the mission control, which applies an hierarchical decision-tree to result in an action. The more detailed information can be found in [Kuhnert & Seemann, 2007][Kuhnert, 2008][Kuhnert et al., 2012].

In this thesis, we emphasise the use of following sensors/devices: CMOS/CCD sensor, PMD sensor, MultiCam, LiDAR, IMU, Stereoscopic imaging cameras, which are mounted on the AMOR in different configurations depending on the requirements of different applications. **Fig. 2.2** shows the configuration of the AMOR when the 2D/3D coupled system is mounted at the front. The system consists of a LiDAR and CMOS camera positioned 5 cm lower. **Fig. 2.3** illustrates the AMOR system when Tyzx DeepSea G2 Camera is used instead of a standard traditional stereo vision

2. FUNDAMENTALS

system. **Fig. 2.4** describes the AMOR's configuration when multiple sensors/devices used simultaneously to evaluate the performances of different approaches, including stereo vision system, LiDAR, CMOS/CCD camera, and MultiCAM.

Figure 2.2: Autonomous mobile robot with LiDAR and CMOS camera mounted near each other to form a 2D/3D coupled system.

2.2 Light Detection And Ranging (LiDAR)

Light detection and ranging is an optical remote sensing technology that illuminates light to a target to measure the distance to or other properties of the target. In this particular section, we would like to discuss about distance measurement using a LiDAR. There are three common laser range-finding techniques: triangulation, pulse time-of-flight and phase-shift measurement.

2.2. LiDAR

Figure 2.3: Autonomous mobile robot with TYZX DeepSea Camera mounted at the front for 3D scene visualisation.

2.2.1 Optical Triangulation for 3D Digitizing

The original idea behind triangulation is to estimate the distance to a target based on the known baseline and angles of emitted and received light. Indeed, assumed that we have the known baseline B, angle of laser source a_1, angle of sensor a_2, the laser source P_1, the sensor P_2, the target P_3, see **Fig. 2.5**. The distance from the laser source to the target is calculated as follow,

$$L = B\frac{sin(a_2)}{sin(a_3)} = B\frac{sin(a_2)}{sin(a_1+a_2)} \tag{2.1}$$

This is quite simple in theory, but turns out to be very challenging in the real world due to the difficulty in measuring the baseline and the angles a_1 and a_2 robustly and repeatably. Therefore, a reliable method is to mount a CCD camera together with the laser system, so that the reflected laser beam passes through the optical axis of the camera.

2.2. LIDAR

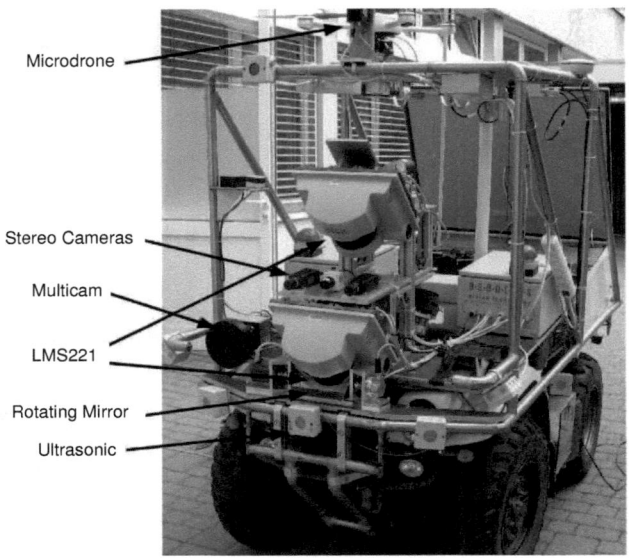

Figure 2.4: Autonomous mobile robot with LiDAR, CMOS camera, and MultiCam mounted at the front up.

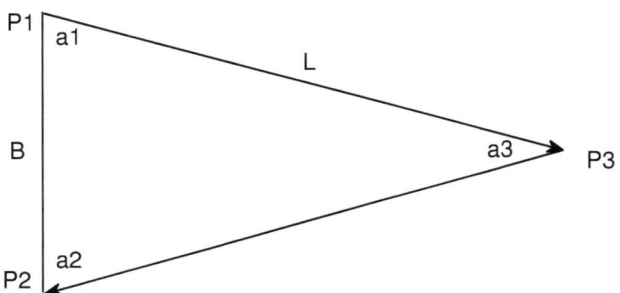

Figure 2.5: Triangulation Configuration

Assume that we have a laser range finder with the configuration as in **Fig. 2.6**. It is quite straightforward to calibrate the system by manually placing a target at the

2.2. LiDAR

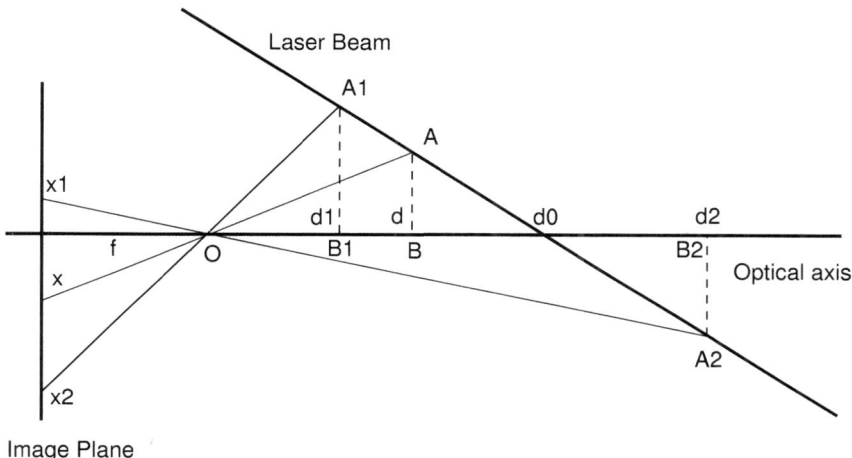

Figure 2.6: Geometric Model

position B_1 with the known distance d_1, and then B_2 with the known distance d_2 in the real world; accordingly determining the distance x_1 and x_2 in the image plane. At an arbitrary distance d in the real world, we firstly determine the distance x in the image plane.

Regarding the property of two similar triangles, we have:

$$\frac{d_1}{f} = \frac{A_1B_1}{x_1}; \quad \frac{d_2}{f} = \frac{A_2B_2}{x_2}; \quad \frac{d_0 - d_1}{d_2 - d_0} = \frac{A_1B_1}{A_2B_2}$$

Thus, d_0 can be derived as follows

$$d_0 = \frac{d_1 d_2 (x_1 + x_2)}{d_1 x_1 + d_2 x_2} \tag{2.2}$$

Based on the property of two similar triangles, we also have:

$$\frac{d_1}{f} = \frac{A_1B_1}{x_1}; \quad \frac{d}{f} = \frac{AB}{x}; \quad \frac{d_0 - d}{d_0 - d_1} = \frac{AB}{A_1B_1};$$

2.2. LIDAR

Hence

$$d = \frac{d_0 d_1 x_1}{d_1 x_1 + d_0 x - d_1 x} \quad (2.3)$$

The distance d can be computed through **Eq. 2.2** and **Eq. 2.3**. In practice, due to the scattering of laser beam, the spatial coherence of the laser light is lost. which means that the depth of eld used at the projection can be useful only if the lens aperture is closed down at the collection. Otherwise the focused laser spot is imaged as a blurred disk of light on the photodetector [Amann et al., 2001]. A solution to this problem is to modify the conventional imaging geometry to conform to the Scheimpug condition, see explanation in [Merklinger, 1996]. Still there exists another problem in the sampling process in the Z axis, which usually requires image pattern centroid location and interpolation; whereby coherence shows its limitation.

2.2.2 Laser Pulse Time-of-flight

The principle behind the Time-of-Flight (ToF) technique for distance measurement is to estimate the amount of time, t, an laser pulse takes to hit the object, be reflected and reach back to the detector. The distance d is then computed as follows.

$$d = \frac{t \times c}{2} \quad (2.4)$$

where c denotes the speed of light and t is the amount of time for the round-trip between the laser and destination. Note that for an unambiguous measurement t should be greater than the pulse width T_p [Jain, 2003].

$$t > T_p$$

or

$$d > \frac{1}{2} c T_p \quad (2.5)$$

While the velocity of light is approximately constant $c = 3 \times 10^8 m/s$. The dis-

2.2. LiDAR

tance d is proportional to the estimated time t. Thus, approximately the error in distance estimation is

$$\delta d = \frac{\delta t \times c}{2}$$

Clearly, the main problem in designing such LiDAR is the realization of an exact time measuring process. This is because the accuracy of the LiDAR depends on the speed of detector and timing circuit used in the device. There exists some sources of inaccuracy in this type of laser range-nders, including noise-generated timing jitter, walk, non-linearity and drift. However, the final precision of distance measurement can be greatly improved by averaging, with the improvement being proportional to the square root of the number of results averaged [Amann et al., 2001]. For instances, the final resolution can be improved to the millimetre level by averaging 100 successive measurements. The main advantage of this technique is its large unambiguous distance measurement which requires a high dynamic receiver with a large bandwidth [Luan, 2001]. Basically TOF laser range finder estimates the distance to a target point by projecting a laser pulse to it. In order to scan many points, often a rotating mirror is used, so that points in a plane can be scanned by sweeping a laser beam horizontally. In this case, the ToF laser range finder is only able to scan in a horizontal direction, the so-called 2D laser scanner. To obtain range information in 3D volumes, another mechanical module has to be added to rotate the scanning module in vertical direction at regular time intervals, the so-called 3D laser scanner [Surmann et al., 2001]. Regarding to 3D model reconstruction, using 3D laser scanner is really time consuming due to the long acquisition time and sometime for converting from point cloud to Cartesian coordinate (because the point cloud is not directly usable). In return, the precision given by that 3D scanner is really precise compared with other 3D scanning devices.

2.2.3 Laser Phase-Shift Range Finder

The idea behind the laser phase-shift range finder is that when modulating the optical power by a constant, the phase-shift between the sent light and the reflected

2.2. LIDAR

light is proportional to the time interval: $\Delta\phi = 2\pi f_m \Delta t$ where f_m is the modulation frequency, c is the speed of light in free space. Hence:

$$d = \frac{1}{2}c\Delta t = \frac{1}{2}c\frac{\Delta\phi}{2\pi}f_m \qquad (2.6)$$

So instead of estimating the time-of-flight of the travelling light, the distance can be deduced from the phase-shift as in **E.q 2.6**. To ameliorate the accuracy of such phase-shift laser range finder, the phase-shift is not directly measured at the working high frequency but at an intermediate frequency $f_i f = |f_m - f_{ol}|$ using a heterodyne technique that preserves the phase-shift versus distance. The limitations of this technique include high level of the photoelectric signal, intermediate frequency drift and electrical crosstalk, see more in [Amann et al., 2001].

2.2.4 Laser Scanner SICK LMS221

As SICK LMSxxx is the most common ToF laser range finder series in industry, we choose the SICK LMS221 which shows impressive performance with very high precision in range mesurement and stability in different environments, please see the default settings of LMS221 as follow.

Parameters	LMS221	LMS221	LMS221
Angular resolution	0.5^o	0.5^o	0.5^o
Aperture angle	180°	100°	90°
Measured Range	$80\ m$	$80\ m$	$80\ m$
Measured value resolution	$100\ mm$	$100\ mm$	$10\ mm$

Table 2.1: Data Sheet of SICK LMS221

From the datasheet, it is seen that the scanner does a 90 degree sweep of the beam with 0.5 degree angular resolution every $10ms$, with the precision is about $10mm$. This gives a scan rate of approximately 18000 points per second.

2.3 Structured Light

Structured light is an active illumination of the scene with specially designed 2D spatially varying intensity pattern. Structured light 3D scanning is about determining the 3D structure of a scene based on the distortion of the projected pattern. In the structured light approach, a light projector and a camera are used. The projector illuminates the scene with a light pattern, and the reflection is captured by the camera, the so-called pattern image. Although many other variants of structured light projection are possible, patterns of parallel stripes are widely used (in black-white or colours). By determining the correspondence between what the projector "sees" and what the camera sees, allows to triangulate the position of every projected pixel and compute its depth. So, it is based on the same principle of passive stereo vision. However, the identification in the structured light approach contrary to the correspondence problem in the stereo case, is easier because the laser spots are normally brighter than the other points in the pattern image which can be identified obviously [Haindl & Zid, 2007]. There usually happens that more than one light plane is projected at a time, which challenges the identification of the light planes. This problem can be solved by encoding the light planes with different indentifications, for example by assigning each light plane a specific colour, the light planes can then be decoded in the pattern image [Forster et al., 2001]. The main drawbacks of such the structured light technology are its strong constraints from which the structured light system can operate properly, such as good scene reflectivity, low contrast of the texture in the scene, see more in [Fechteler & Eisert, 2008].

2.4 The MultiCam

The MultiCam (see **Fig.** 2.7) is actually integrated from a CMOS camera and a Photo Mixer Device (PMD) camera. The MultiCam consists of two imaging sensors (a conventional 10-bit CMOS sensor with VGA resolution and a PMD sensor with 3K resolution), a dichroic beam splitter, a near-infrared light system, FPGA based processing unit and USB 2.0 communication interface [Ghobadi et al., 2010]. A

2. FUNDAMENTALS

general optical set-up can be seen in **Fig. 2.8**.

Figure 2.7: MultiCam

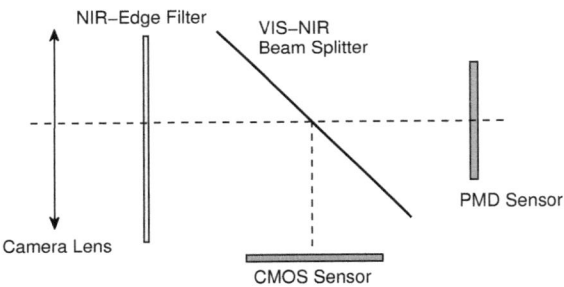

Figure 2.8: Optical setup of the MultiCam

A single lens is used to gather the light for both sensors, so a 2D-3D calibration is not necessary but an easy image registration can be done by a two dimensional translation function which maps a 10x10 2D pixel to one single PMD pixel. While the 3D sensor needs to acquire the modulated near-infrared light (about 870 nm) back from the scene, the 2D sensor is used to capture the images in the visible spectrum (approximately from 400 nm to 800 nm). As a result, the MultiCam provides simultaneously four images: including depth image (64x48 pixels), modulation image (64x48 pixels), NIR intensity image (64x48 pixels) and colour image (640x480 pixels) [Möller et al., 2005]. The colour image is simply obtained from the CMOS sensor. In order to understand how other images are generated, we will summarize the principle operation of the PMD camera which uses Time-of-Flight (TOF)

2. Fundamentals

technique to measure distances. The principle of the range measurement in a TOF camera, similar to a laser range finder, is based on the measurement of the time the light needs to travel from one point to another. This time which is so-called Time-of-Flight is directly proportional to the distance the light travels because the velocity of the light is approximately constant at $3^8 m/s$. However, a direct measure of the time difference for each single emitting ray is infeasible, thus, a frequency modulation process is applied for the active light source. Assume that, we use continuous sinusoidal modulation at frequency f_{mod}. The phase shift can be calculated in terms of time: $\Delta \varphi = 2\pi . f_{mod}.t$. Besides, if we take four samples A_1, A_2, A_3, and A_4 each shifted 90 degrees, the phase-shift of the sent and received signals can be computed as

$$\Delta \varphi = arctan(\frac{A_1 - A_3}{A_2 - A_4}) \qquad (2.7)$$

Hence, the distance is calculated as follows [Möller et al., 2005]:

$$d = \frac{c.\Delta \varphi}{4\pi . f_{mod}} \qquad (2.8)$$

The strength of the received signal a and the NIR intensity information b are expressed as [Möller et al., 2005]:

$$a = \frac{sqrt(A_1 - A_3)^2 + (A_2 - A_4)^2}{2} \qquad (2.9)$$

$$b = \frac{A_1 + A_2 + A_3 + A_4}{4} \qquad (2.10)$$

The high frame rates can be achieved at around 50 fps to 60 fps, which is comparable to a regular video camera. Thus, the MultiCam is suited for a real-time application.

Limitations

- As mentioned above that the MultiCam uses a modulated lighting system so

2. FUNDAMENTALS

the unambiguous range measurement in the camera is restricted. For example if the modulation frequency is at $20MHz$, it is limited to $7.5m$. While the objects over this distance can be observed in 2D image of the MultiCam, they do not have any reliable distance information in the 3D image. Although reducing the frequency can increase the unambiguity of range measurement, it reduces the resolution of range measurement as well (see more in [Ghobadi et al., 2010]).

- The work of [Nguyen et al., 2010a] proposed a distance compensation to extend the range of measurable distances, the performance was not impressive as well as stable in different lighting conditions.

- For a good depth perception of the scene, a powerful lighting system is required, which might increase the cost.

- The affection of the sunlight is huge to the distance measurement results.

Those limitations defeats outdoor applications of the MultiCam concerning distance measurement. Nevertheless, in our work, we mainly do not use the distance information but NIR intensity and colour information. Also taking into account the energy reduction of the modulated light during its travel, we restrict the maximum distance at $50m$. The aim is to obtain the reflected light with strong energy enough to be classified with other NIR light from the sunlight or other light reflectance sources. **Fig. 2.9** illustrates some image samples captured by the MultiCam.

2.5 Stereoscopic Imaging

Stereoscopic imaging is a passive triangulation method, so it does not require any light sources, but multiple 2D sensors aligned. In a classical stereoscopic vision technique, the so-called stereo vision, two cameras are employed in a binocular vision system, analogous to the two eyes in the human visual system, to obtain two differing views on a scene. The idea is simple that the correspondence between the two views

2. Fundamentals

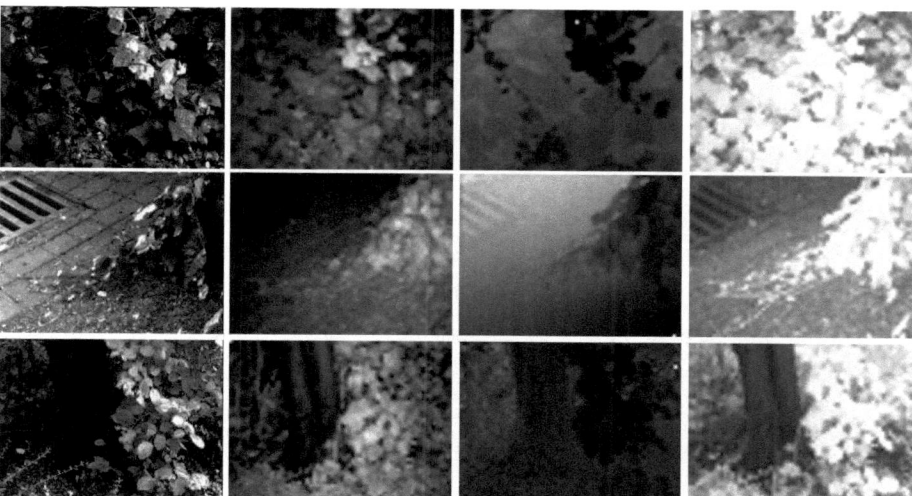

Figure 2.9: Examples of MultiCam's images (from left to right): 2d; modulation; depth; infrared intensity. Those images were captured around the campus Hölderlin of Universität Siegen.

are matched, then the depth can be estimated based on triangulation by knowing the camera focal lengths. In reality, the task of robustly finding the correspondence is challenging due to the imperfection of lens, low quality of stereo-pair, illumination noise and illumination effects. Thus, there are some pre-processing steps needed to do beforehand such as distortion removal and image rectification. In general, there are three main steps in any stereo vision techniques: Calibration, Rectification, and Stereo Matching.

Calibration: is to find the mathematical transformation that converts the 2D coordinates of pixels in the stereo images into the real world 3D coordinate. For that aim, each camera is self-calibrated to obtain the internal information or intrinsic camera parameters. Determining the correspondence of the stereo images help to solve the camera calibration problem to achieve exterior information or extrinsic camera parameters. The formation resulted from both intrinsic and extrinsic parameters is the transformation matrix of the stereo system.

2. FUNDAMENTALS

Rectification: is the process of re-sampling stereo images so that the epipolar lines correspond to image rows. The basic idea is simple that if the left and right image planes are coplanar and the horizontal axes are co-linear (no rotation about the optical axes), then the image rows are epipolar lines and stereo correspondences can be found by searching for matches along corresponding rows. In practice, this condition can be difficult to achieve and some vergence (inward rotation about the vertical camera axes) may be desirable, but if the pixels in the left and right images are projected onto a common plane, then the ideal epipolar geometry is achieved.

Stereo Matching: combines the two images obtained from the rectification process and takes the position of the pixels in the left image to output the corresponding pixel location in the right image. With this method we calculate the pixel's distance from the camera. The depth is then translated to a depth map where points closer to the camera are almost white whereas points further away are almost black. Points in between are shown in gray-scale, which get darker the further away the point gets from the camera, see **Fig. 2.10** for an example depth map with the original image. If there is no change to the configuration of the stereo system, the first two steps only need be done off-line once, whereby the returned parameters can be used for the online stereo matching process. Thus, the speed is improved.

The fact is that the stereoscopic imaging technique requires a computational software in order to result in 3D scenes from pairs of images. A clever idea to improve the speed of the process is to convert the software into hardware, given by Tyzx company. In that way, a block matching algorithm used for stereo calibration is fast implemented in just around 10 ms for an image resolution of 740x468 pixels. Consequently, the frame rate for depth image acquisition is about 60 fps instead of 3 fps as in the software approach, which is really impressive.

Still, the main drawback of stereoscopic imaging approach is that no range data can be obtained in uniform regions, like a white wall, where there are no features present for the correspondence process [Sobottka, 2000]. The shadowing effect is also a typical problem for stereo vision systems which can be minimized by using multi view triangulation systems at the price of an enormous increase of data processing as well as increasing the number of cameras. Finally, illumination effects

2.5. Stereoscopic Imaging

Figure 2.10: Depthmaps (the second row) with the corresponding pictures (the first row), gray values show the depth of the images. Those images were captured around the two campuses Hölderlin and Paul-Bonatz of Universität Siegen.

cause much noise in the acquired depth images, which restricts the applicability of stereoscopic imaging techniques.

The next subsection will contribute a fitting plane algorithm-based depth correction for Tyzx stereoscopic imaging where problems of illumination noise and no range data in uniform regions are completely solved.

2.5.1 Fitting Plane Algorithm-based Depth Correction for Tyzx DeepSea Stereoscopic Imaging

The work presented in this paper deals with the poor performance of depth image generation given by a Tyzx stereo vision system under different lighting conditions

2.5. FITTING PLANE ALGORITHM-BASED DEPTH CORRECTION

in both indoor and outdoor environments. For that aim, we introduce a fitting plane algorithm to correct distance information as well as fulfil the missing points in the original depth. First, the colour image is over-segmented into many small homogeneous regions of interest. Those small regions can be approximately considered as planar surfaces which form the 3D scene. While 3D points inside each small region should found a plane, this insight is then used to enhance the depth image. Assuming that the environment is made up of a number of small planes, we certainly make no explicit assumptions about the structure of the scene; this enables the algorithm to cope up with many different scenes even with significant non-vertical structure.

The algorithm has been confirmed to be easily implemented and robust throughout many experiments in different lighting conditions and different scenarios in both indoor and outdoor environments. Concretely, the proposed approach enables a 3D reconstruction capability using Tyzx DeepSea G3 vision system which is infeasible from the raw depth data. Moreover, the proposed algorithm improves more than 48% of 3D reconstruction accuracy compared with the original result given by the stereo vision system over testing 611 scenes under real-time constraint.

This work has been published in **Proceeding of ICCE-2012** [Nguyen *et al.*, 2012a].

2.5.1.1 Introduction

Reconstructing 3D environments is one of the most popular research areas in computer vision and computer graphics, it is widely used in many fields, such as animation, video game, robotics, and so on. Commonly 3D reconstruction techniques can be divided into two categories: active and passive. Active approaches based on light structure [Forster *et al.*, 2001][Ohta, 2007], laser range finder [Surmann *et al.*, 2001][Jain, 2003] and time-of-flight [PMD, 2009] can directly provide 3D information, which nevertheless could not be used in some circumstances due to limitations of active source properties(using modulated light can be strongly affected by the sunlight; using laser costs too much time for data acquisition; etc.) and low resolution. Moreover, for many purposes, researchers still need to use colour information

2.5. Fitting Plane Algorithm-based Depth Correction

which is not available under these approaches. In contrast, stereoscopic imaging is a passive triangulation method in which depth information about a scene is measured from multiple static 2D images, each acquired from a different viewpoint in space.

Given the stereo geometry, the 3D image of the scene can be reconstructed after a computational process of affine transform. However, it is computational expensive for a very accurate searching and matching process. Furthermore, no range data can be obtained in uniform regions, like a white wall, where there are no features present for the correspondence process.

First, there must be no doubt about how important the speed of stereo vision process is, with respect to real applications. Indeed, a fast and robust stereo vision system is able to simulate what human eye sees in real time, the outcome as an on-line 3D model of the viewed scene can lead to ease many autonomous tasks such as obstacle avoidance, terrain classification, object tracking, object detection and recognition, which are often used in autonomous navigation or man-machine interaction. There are two ways to speed up the real-time stereo vision, the first way is to parallel the algorithms for stereo vision [van Beek & Lukkien, 1996]. The other way is to use some hardware to speed up. Concretely, in this work, we present the Tyzx DeepSea G3 Stereo Vision System which includes a stereo camera, on-board image rectification, and an interface to a general purpose processor over a PCI bus [Woodfill et al., 2004]. The system is based on the DeepSea processor which computes the depth based on simultaneously captured left and right images with high frame rate. The chip can run at 200 frames per second with 740x468 images. Due to some other hardware issues, the speed should be slowed down to maximum 60 frames per second to obtain a good result. Regarding to the field of robotics, we are really satisfied with that fast system, please see characteristics of elements of the camera in **Table 2.2**.

The remaining issue as said is how to cope up with the case of uniform regions, point missing in the raw depth, and noise caused by illumination effects. This is also the aim of this paper. In order to understand our solution, let's turn back to the idea of how openGL and DirectX models used to build 3D scenes. Actually they use triangular facets to model shapes, even very complex shapes. Therefore, we start

2.5. FITTING PLANE ALGORITHM-BASED DEPTH CORRECTION

Table 2.2: Characteristics of Elements

Dimensions	3.8cm.7cm.5cm
Temperature Range	$-40^{\circ}C$ to $+85^{\circ}C$
Weight	675 g
Power	12W typ. 12 vdc or PoE class III
Frame rate	60 FPS
Image size	740x468
Lens options	40°, 62°, 83° Horizontal FOV
Baseline options	3cm, 6cm, 8cm, 14cm
Stereo Algorithm	Census
Search	64 Disparity + 4 bit subpixel
Pixel	10 bit or 12 bit
CPU	PowerPC 64 bit data bus
Memory	256 MBytes
Operating System	Linux 2.6 Kernel

with the idea of dividing the viewed scene into many small regions. Thus, each small region should be a planar surface. An algorithm to find the best fit plane to describe the planar surface of each region is introduced in this paper. The distances between 3D points inside the region to the plane can either tell us about the smoothness of the surface or which points seem to be wrong measured, thus, need to be corrected. Our algorithm was able to automatically enhance depth images that were both qualitatively correct and visual pleasing for 611 pairs of test images with more than 48% improved in 3D reconstruction accuracy compared with the raw depth data, see an example in **Fig. 2.11**. Additionally, we also prove that good depth results can be obtained based on our approach in real-time.

The rest of this paper is organized as follows. Subsection 2.5.1.2 discusses the intuitions from human vision to stereo vision in order to establish the fitting plane. Subsection 2.5.1.3 presents the fitting plane algorithm. Subsection 2.5.1.4 illustrated experiments and results. Finally Subsection 2.5.1.5 concludes this work.

2.5. Fitting Plane Algorithm-based Depth Correction

Figure 2.11: (a) 2D image. (b) original depth (Best viewed in colours: orange(near); green(neutral); purple(far); white(very far). In the same colour: the darker the nearer). (c) corrected depth by proposed algorithm. (d) 3D scene reconstructed

2.5.1.2 Planar Surface for Scene Understanding

Given a 2D image, human eyes use many monocular cues to infer the 3D structure of the scene. The cues are formed by firstly separating the image into many small pieces and then together with geometrical intuition to imagine the projection for those pieces(or : cue = piece + projection). For example, asking a kid to build a 3D scene from a still single 2D image, he/she would prefer to cut the image into many small parts, then re-arrange them based on his/her geometrical intuition about the scene (see **Fig. 2.12-(Middle))**. This proves one thing that 3D structure can be intuitively modelled as a formation of many different planar surfaces.

In order to build the planar surfaces, we first need to segment image into many small regions. Using superpixel image segmentation technique [Felzenszwalb & Huttenlocher, 2004], an example of over-segmented image is shown in **Fig. 2.12**. The reason to use such segmentation technique is because it provides a relative good

2.5. FITTING PLANE ALGORITHM-BASED DEPTH CORRECTION

Figure 2.12: **Left**: An image of a scene. **Middle**: Simple cuts to construct 3D scene from one single 2D image. **Right**: over-segmented image where each small region (superpixel) lies on a plane in the 3D world.

segmentation (neither too coarse nor too fine, see more in [Felzenszwalb & Huttenlocher, 2004][Saxena *et al.*, 2009]). Additionally, the computation is not expensive compared with other segmentation techniques (implemented to run in $O(mlogm)$ time).

Looking into more details of the over-segmented image which contains many small regions of interest, each region is approximately homogeneous in colour. A good observation one can recognize that a uniform region is usually extracted in a larger size compared with others containing textures and edges. A carefully reading one must raise the question of how to determine the depth of those uniform regions in which there is no depth information obtained from the stereoscopic imaging process. We will answer that question later on when we have already corrected depth information in areas which partially have raw depth data. Seeking the solution that fulfils the missing points and corrects wrong or noise ones in the depth image, we first start with regions which contain texture or have depth information from the raw depth data given by the Tyzx DeepSea vision system.

So far, we continues now to understand how to build a planar surface for each small region which has depth information in several scattered 3D points. Assume that a small region has N pixels in the colour image but only M 3D points in the depth image ($M \leq N$). Thus, we have $N - M$ missing points. Ideally all the points

2.5. Fitting Plane Algorithm-based Depth Correction

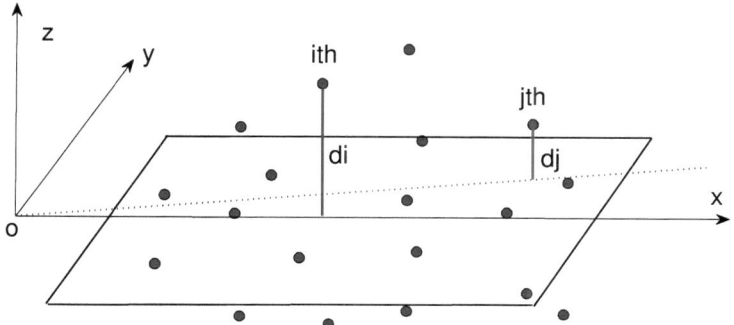

Figure 2.13: A best fit plane for a set of given 3D points.

lies on the same plane ax + by + cz = d, thus, we have

$$a \times x_i + b \times y_i + c \times z_i = d \qquad (2.11)$$

In fact, the real surface is not completely smooth as well as affection from noise and wrong measurement, thus, **Eq. 2.11** is not correct for all 3D points $(x_i, y_i, z_i) \forall i = \{1 \rightarrow M\}$. This turns out to an optimization problem of finding the best fit plane where the sum of distances from the M 3D points to the plane is minimum, see **Fig. 2.13**. Or, we have to find the variables (a,b,c,d) so that the distance α is minimum.

$$\alpha = argmin(D(a,b,c,d))_{\{a,b,c,d\} \in \Re; (a \times b \times c \times d) \neq 0} \qquad (2.12)$$

where

$$D(a,b,c,d) = \frac{\sum_i (ax_i + by_i + cz_i + d)^2}{a^2 + b^2 + c^2} \qquad (2.13)$$

If we can solve the optimization problem in **Eq. 2.12**, missing points of the region in the depth image can be fulfilled. For example, the depth information z_j of a missing point at the position (x_j, y_j) in the depth image can be estimated as

$$\widehat{z_j} = \frac{\widehat{d} - \widehat{a} \times x_j - \widehat{b} \times y_j}{\widehat{c}} \qquad (2.14)$$

2.5. FITTING PLANE ALGORITHM-BASED DEPTH CORRECTION

$(\hat{a}, \hat{b}, \hat{c}, \hat{d})$ are estimation values of (a, b, c, d) after the optimization process. Even we do not know exactly what structure type of the cube corresponding with a small region, it is still true to assume that every small region should represent a quite smooth surface. Therefore, if we calculate the average of distances from the M 3D points to the fitting plane, we expect that all correct points should have the distances smaller than three times of the average, or

$$d_k \leq d_{avg} = 3 \times \frac{\sum_{i=1}^{M} d_i}{M}; \forall k = \{1 \to M\} \qquad (2.15)$$

A point O is considered as a *defect* point if and only if O belongs the small region and $d_o > 3 \times d_{avg}$. In that case, a repetition of the optimization process needs to be carried out again without the point O. O is then treated as a missing point.

Solving the optimization problem

Regarding to **Eq. 2.13**, If we set the partial derivative with respect to d equal to zero, we can solve for d to get

$$d = -(a \times x_c + b \times y_c + c \times z_c) \qquad (2.16)$$

where (x_c, y_c, z_c) is the centroid of the points. So, we are finding a plane which pass through the centroid and has least square distance to all points in the region. If we substitute it back into **E.q 2.11** we get

$$a \times (x_i - x_c) + b \times (y_i - y_c) + c \times (z_i - z_c) = 0 \qquad (2.17)$$

If we define the vector $A^T = [a\ b\ c]$ and $X = \{X_i : X_i = [x_i\ y_i\ z_i], \forall i \in \{1 \to M\}\}$. Eq. 2.13 can be re-written as

$$D(a, b, c) = \frac{A^T X^T X A}{A^T A} \qquad (2.18)$$

Let $Cov = \frac{1}{M} X^T X$ is covariance matrix of the data, so the distance $D(a, b, c)$ is a

2.5. Fitting Plane Algorithm-based Depth Correction

Rayleigh Quotient which is minimized by the eigenvector of X that corresponds to its smallest eigenvalue. Therefore, we simply find the eigenvalues and eigenvectors of *Cov* by Singular Value Decomposition, the eigenvalues of *Cov* are the squares of the singular values of X, and the eigenvectors of *Cov* are the singular vectors of X. Then the smallest sum distance is equal to the smallest eigenvalue of *Cov*. The three eigenvectors are mutually orthogonal and define three sets of (a,b,c). Thus, we want to choose the eigenvector associated with the smallest eigenvalue. The optimization problem has been solved.

Depth estimation for uniform regions

In fact, it is possible to use Markov Random Field (MRF) to model the depth information of uniform regions. That means the depth information of a uniform region only depends on the depth information of the region's neighbours. However, this MRF needs some times for training and evaluating, thus, destroys the real-time constraint. Consequently, this work investigates another way for depth estimation that can be fast implemented but still tolerates qualitative performance as well as visual pleasing. Consider an example of uniform region O in **Fig. 2.14**, there are only few depth points in that region while we also are not sure if those points are correct or not. A good observation in the over-segmented image in **Fig. 2.14** can point out the following insight:

Property 1: There are no two adjacent uniform regions. In other words, a uniform region should connect with many other non-uniform regions.

Proof: An over-segmentation process segments the 2D image into many small pieces where edges are also segmented as small regions. Consequently, the edges of each uniform region are also existed in form of region of interest. Therefore, the neighbours of the uniform regions are edges regions or small textured regions, or there are no two adjacent uniform regions.

This enables an idea of estimating depth information of the uniform region by establishing another fitting plane which has minimum sum distance to all 3D points of neighbor regions. So, we actually turn back to the solved problem of finding a fitting plane for a set of 3D points.

2.5. FITTING PLANE ALGORITHM-BASED DEPTH CORRECTION

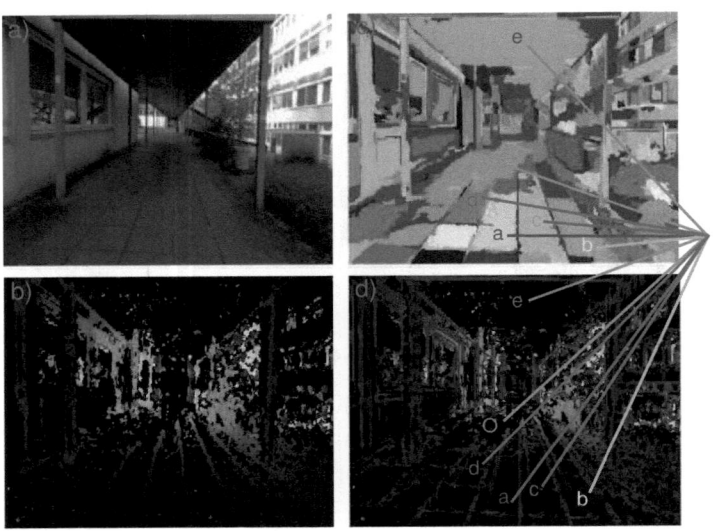

Figure 2.14: a) Gray-scale image. b) Raw depth (Best viewed in colour, the colour code is orange: near; green: far; purple: very far, for each colour: the darker the nearer). c) Over-segmented image. d) Mapping regions of interest where the contours of segmented regions are marked in blue colour.

2.5.1.3 Fitting Plane Algorithm

Algorithm ────────────

Step 1: Segment 2D image into many small regions using superpixel technique.

Step 2: Find textured regions which should contain a significant depth information (in our case the regions have more than 10% depth information). Repeat step 3 and step 4 for all textured regions.

Step 3: Establishing the fitting plane for a set of M 3D points obtained from raw depth data.

Step 4: Comparing the distance of each 3D point to the fitting plane.

- If there are m 3D points ($m > 0$) whose distances are superior than three times

2.5. Fitting Plane Algorithm-based Depth Correction

of the average, treat the *m* points as missing points. Repeat the step 3 with $M - m$ 3D points.

- if $m = 0$. All missing points in the depth image can be fulfilled based on the **E.q 2.14**.

Step 5. For each non-textured region or uniform region, search for all 3D points of the neighbour regions in the depth image and consider them as the initial depth information of the uniform region. Repeat step 3 and step 4 for all uniform regions.

Regarding to the time issue and illumination noise, it is better to downsample 2D image using Gaussian Pyramid before taking the segmentation process. On the other hand, looking at the raw depth data in **Fig. 2.15** and **Fig. 2.16**, there exists many noise or errors (appeared as white spots) which causes the repetition of step 3 when runing the step 4 (see the first option *if* in the step 4), thus, slowdown the speed of the algorithm's implementation. A simple way to overcome this issue is to take a pre-processing step for depth denoise. We suggest to use a continuity property of a planar surface that means the difference in distance of two continuous points should be smaller than the difference from each point to the centre of the plane. In order to realize the idea, from mathematic point of view, we can simply **SORT** the distance values of all points in a small region, so that $R_i = \{P_j : P_j < P_{j+1}, j = \{1 \rightarrow M\}\}$.

Figure 2.15: (Left) Raw depth. (Right) Depth refined.

2.5. FITTING PLANE ALGORITHM-BASED DEPTH CORRECTION

The continuity is expressed through:

$$\begin{cases} P_j - P_{j+1} < P_j - P_c \\ P_j - P_{j+1} < P_{j+1} - P_c \end{cases} \qquad (2.19)$$

$\forall j = \{1 \rightarrow M-1\}$ where P_c is the centre point of the plane. If **Eq. 2.19** is not satisfied, then the point P_{j+1} is refined as $P_{j+1} = P_j$.

An example of the refinement result is illustrated in **Fig. 2.15**. Then, we will gain the faster implementation of image segmentation due to the smaller size as well eliminate illumination noise. Generally, the segmentation process takes around 40ms to 60 ms depending on the complexity of the image texture. The whole fitting plane algorithm runs at around 346 ms, so the frame rate is at about 2.8 fps (with CPU 2.4 GHz, 4G RAM).

2.5.1.4 Experiments and Results

In order to evaluate the proposed algorithm, we did two main experiments. First, we used Tyzx DeepSea G3 vision system to take many images in indoor environments such as in office and corridor. Second, we mounted the Tyzx DeepSea G3 vision system in front of our autonomous mobile robot to collect data when the robot traversed throughout outdoor environments, see **Fig. 2.3** in section 2.1. Consequently, 200 indoor scenes and 311 outdoor scenes were captured and used to evaluate the performance of the given approach. The good thing is that the way of segmenting 2D images into regions of interest also helps to devise a way for precision measurement. That is to count the percentage of segments to be correctly reconstructed. Several examples of depth corrected by the proposed algorithm are shown in **Fig. 2.16**. Intuitively, the results are visual pleasing and demonstrate a significant improvement to the depth information obtained by Tyzx DeepSea G3 vision system. Clearly the depths of the uniform regions have been reasonably estimated as well as depth defects are mostly eliminated for scenes in different scenarios and different lighting conditions. **Table 2.3** describes the accuracy of depth correction through counting the number of corrected facets reconstructed. The comparison between our algo-

2.5. Fitting Plane Algorithm-based Depth Correction

rithm performance and raw depth data is described in **Table 2.4** where the proposed approach improves 48.22% of true depth to the raw result.

Table 2.3: Depth Correction Accuracy

	indoor	outdoor
No. scenes	200	311
No. facets	2000	3110
True depth (%)	68.41	72.83

Table 2.4: Comparison

	Raw Depth	Corrected Depth
No. scenes	611	611
No. facets	6110	6110
True depth in average (%)	71.06	22.83

Figure 2.16: The first row describes 2D images. The second row show the corresponding raw depth data. The last row demonstrates the depth correction given the proposed algorithm (Best viewed in colours: orange(near); green(neutral); purple(far); white(very far). In the same colour: the darker the nearer).

2. FUNDAMENTALS

2.5.1.5 Conclusion

We have introduced the fitting plane algorithm for depth correction to enable 3D reconstruction capability using Tyzx DeepSea G3 vision system in both indoor and outdoor environments. Experiments and results demonstrate that the proposed algorithm provides a robust depth correction in different scenarios: from an urban scene where the main structures are linear and smooth surfaces (line; wall; road) to a forest scene where many textured regions appear (tree; grass; soil). Compared with the raw depth data, the accuracy of true depth reconstruction is improved more than 48% by our approach. The algorithm runs fast at around 2.8 fps, thus, can be used for real-time 3D reconstruction. The limitation of the approach is to deal with very far objects which lack depth information but more affected by illumination noise (that can bee seen in **Fig. 2.16** where the results of depth correction is not that good). The future work should investigate the interaction between neighbour planar surfaces in order to improve the accuracy of depth correction.

2.6 Multi-spectral Imaging

A traditional digital camera is designed to capture the light that falls onto the sensor in a fashion that resembles the human perception of colour. For that aim, wideband filters are used to obtain red (R), green (G), and blue (B) channels. In contrary, multi-spectral imaging enables us to capture information that might be available or unavailable to the human observer. The considered light spectrum might ranges from short-wavelength violet to infrared, depending on different purposes and applications. A multi-spectral image is captured at specific frequencies across the electromagnetic spectrum. In general, there exists two common ways to capture a multi-spectral image. First, the light source is modulated from which the projecting light focuses on a narrow band at a time. Whereby the spectrum of a single point is measured by continuously shifting the band of the modulated light, and the entire eld of view is scanned over time. The common device using this technique is the spectrometer. Second, the basic idea is to separate the incoming light into its spectral components,

2. Fundamentals

which are then sensed by many monochrome 2D sensors. Each of these sensors are only sensitive at a specific band, thus we are able to obtain multiple multi-spectral images at different bands from these different sensors. In fact, the use of multiple 2D sensors is costly and not reliable due to difficulties in hardware design. Thus, only one high dynamic range 2D sensor which is adjustable in its spectral sensitivity is often used. By changing the spectral sensitivity of the sensor over time, multiple multi-spectral images are captured. A popular way of changing the spectral sensitivity is to use prisms, diffraction gratings, gel lters or tunable lters [Gat, 2000]. Existing systems differ in terms of how they trade off spatial and temporal resolution to obtain multi-spectral measurements for each point in the eld of view. Some hybrid approaches can simultaneously scan a static scene with respect to space and spectrum by modifying a commodity camera. High-cost devices that use complex optics and custom photo-sensors have been developed for remote sensing that can acquire hyperspectral videos of dynamic scenes. Overall, all these systems are quite expensive and long data acquisition. Thus, the traditional multi-spectral approaches are often used in the remote sensing field and military applications, but yet not applied in civilian applications especially in case of requiring a real-time constraint.

Interestingly the reflectance of the scene at different bands is represented through the captured multi-spectral images. So, multi-spectral imaging helps to understand image formation and reflectance phenomena. Thus, research on computer vision methods that interpret, or rely on, scene reflectance often profits from analysing those multi-spectral images. However, extracting useful information from multiple multi-spectral images costs much computational effort while the efficiency varies significantly against the illuminating conditions where those images are captured. In the robotics research, multi-spectral images are only taken in several fixed bands in order to reduce the cost of building multi-spectral scanning devices and of computation in multi-spectral image processing. For example, multi-spectral cameras with wide band filters are designed to obtain red, green, blue and near-infrared channel, which are usually used in agricultural applications, especially for detecting and analysing fruits and vegetation. **Fig. 2.17** illustrates samples of multi-spectral images captured by DeepSea Stereo Camera with NIR-Transmitting filter. Besides,

2. FUNDAMENTALS

visible and infrared sensors integrated in a monocular setup like in the MultiCAM 2.4 or in the work of [Bradley *et al.*, 2007] also provide simultaneously red, green, blue and near-infrared information.

Figure 2.17: The first row describes colour images where each image consists of red, green and blue channels. The second row shows the corresponding infrared images.

Chapter 3

Vegetation Indices Applied for Vegetation Detection

For use in the photosynthesis process, chlorophyll, the most well-known and most important pigment causing the green colour of healthy plant leaves, strongly absorbs visible light (from 0.4 to 0.7 μm), especially red and blue light (see **Fig. 3.1**). The cell structure of the leaves, on the other hand, strongly reflects near-infrared light (from 0.7 to 1.1 μm) (see **Fig. 3.2**). The more leaves a plant has, the more these wavelengths of light are affected, respectively. This enables vegetation indices (VIs) which are defined as combinations of surface reflectance at two or more wavelengths designed to highlight a particular property of vegetation. For instance, the ratio of radiances in the near-infrared (NIR) and Red bands has been used as a measure of vegetation index in the satellite remote sensing field [Tarpley et al., 1984] [Wurm et al., 2009] [Crippen, 1990] [Manduchi, 2005]. There are many different vegetation indices devised in order to detect vegetation in different scenarios, which are described in 3.1.

One might raise a question if such vegetation indices are really useful in detecting different species/types of vegetation, whose amount of chlorophyll in their leaves diverges considerably. Intuitively, dying vegetation (usually appeared in yellow, brown or red colour) contains very little chlorophyll. To answer that question, an estimation on reflectance of different types of vegetation from the sunlight was carried out by

3. VEGETATION INDICES

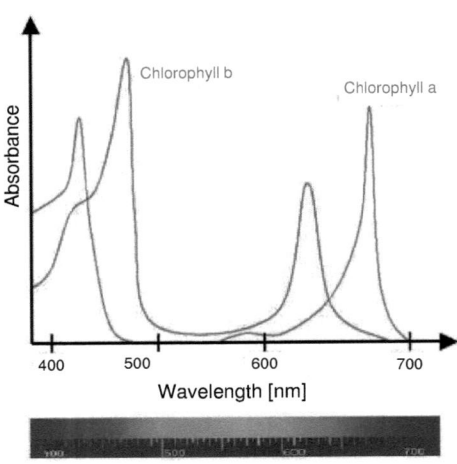

Figure 3.1: Absorbance Spectra of Chlorophyll a (green) and b (Red) [Asner, 1998]

NASA (USA), using NOAA-AVHRR (National Oceanic and Atmospheric Administration - Advanced Very High Resolution Radiometer) instrument. Accordingly, healthy vegetation (**Fig. 3.3 Left**) absorbs most of the visible light that hits it, and reflects a large portion of the near-infrared light. Unhealthy or sparse vegetation (**Fig. 3.3 Right**) reflects more visible light and less near-infrared light. Therefore, it should be made clear that all traditional vegetation index-based approaches tend to detect only chlorophyll-rich vegetation, or green one. Chlorophyll-less vegetation is usually mis-detected or confused with wet soils and other material surfaces.

Even though those vegetation indices have been widely and successfully used in many remote sensing applications, for example classifying and positioning the green areas of the earth surface, it is still a problematic thought to apply them directly for mobile robotics applications due to drastically different view-points. Regarding to autonomous ground navigation, there would be more complications to deal with, such as illumination effects (shadow, shining, under-overexposure), views of sky,

3. Vegetation Indices

and presence of variety of different materials, from which the reflected light can have a spectral distribution that is different from that of the sunlight. This explains why not much investigation is available on utilizing vegetation indices in the field of robotics, except few works done for automatic fruit detection. Remarkably, a quite impressive approach applying vegetation indices for detecting vegetation in autonomous ground vehicles was introduced by [Bradley et al., 2007]. Again, the huge affection from illumination effects restricts the applicability of the approach. [Bradley et al., 2007] then had to additionally use LIDA data and colour information to extract more features in order to strengthen the vector components. This, however, does not meet real-time constraint due to long-time data acquisition of Laser Scanner (2s in average to acquire a frame with 6437 scanned points) [Nguyen et al., 2011b].

In order to be more stable against illumination changes in outdoor environment, and also satisfy the real-time constraint, a new vision system set-up which combines CMOS sensor and Photo Mixer Device sensor with a near-infrared lighting system is introduced to simultaneously provide near-infrared and colour images at high frame

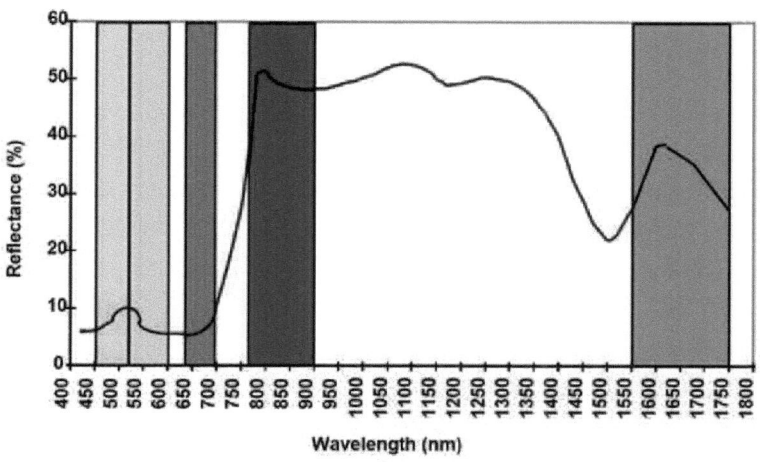

Figure 3.2: Reflectance Spectrum of Green Leaf [Asner, 1998].

3. VEGETATION INDICES

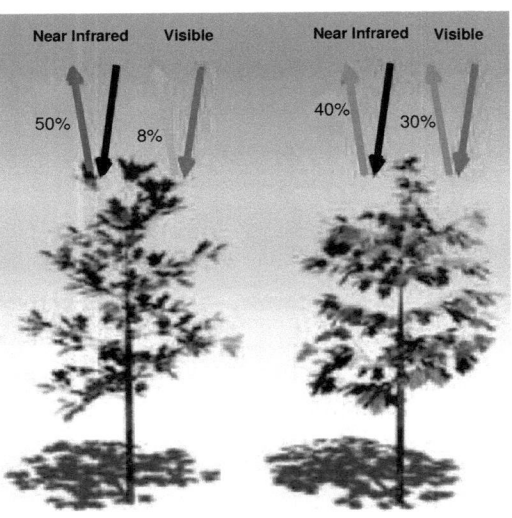

Figure 3.3: Absorption and Reflectance of Green (Left) and Brown (Right) Vegetation [NASA, 2012].

rate. Those near-infrared and colour information are then used to compute a novel vegetation index which is derived from doing regression analysis on NIR and Red reflectance, and luminance data of vegetation pixels. The novel index is so called as Modification of Normalized Difference Vegetation Index (MNDVI) due to its similarity in formulation with the traditional form Normalized Difference Vegetation Index. MNDVI is then defined as the new standard form of vegetation index for such vision system integrated with an additional lighting system. The novel vegetation index is proved to be more stable and efficiently used for detecting vegetation in different lighting conditions and under real-time constraint. More interestingly, empirical evidences demonstrate that MNDVI can help to also detect living chlorophyll-less vegetation (Brown/Yellow/Red colour leaves), which is infeasible in the traditional vegetation indices-based approaches.

3. Vegetation Indices

The chapter is organised as follow: Section 3.1 introduces related works. Section 3.2 presents a novel approach for real-time vegetation detection using an active NIR lighting source. Section 3.3 discusses and compares the performance of the proposed approach with conventional ones while section 3.4 summarises and concludes this work.

This work has been published in **Journal of Robotics and Automation** [Nguyen et al., 2012c].

3.1 Related Work

The spectral properties of chlorophyll-rich vegetation are primarily determined by the absorption spectra of water and chlorophyll, and the refraction of light at cell walls [Willstatter & Stoll, 1913]. The water presents in cells absorbs light with wavelengths longer than 1400 nm. Chlorophyll strongly absorbs visible light, especially red and blue wavelengths [Clark et al., 2003]. The remaining light is efficiently scattered by the critical internal reflection caused by the change in refractive index from water to air at the cell wall. As a result, those wavelengths between 800 nm and 1400 nm that escape both water and chlorophyll are strongly reflected in all directions. Thus, in order to detect vegetation, a simple threshold on vegetation indices should do the work. The following subsections will introduce most common vegetation indices used in both remote sensing field and robotics research.

3.1.1 Ratio Vegetation Index

[Jordan, 1969] assumed that lines of equal vegetation all intersect at the origin and developed the Ratio Vegetation Index.

$$RVI = \frac{NIR}{Red} \quad (3.1)$$

The RVI measures the slope of the line between the origin of Red-NIR space and the Red-NIR value of the pixel. The higher value of RVI a pixel has, the more likely

3.1. RELATED WORK

is it vegetation one. Nevertheless, the absolute value of RVI varies considerably due to light intensity changes, which restricts the applicability of the index in reality.

3.1.2 Normalized Difference Vegetation Index

Although a simple ratio of Band 5 (Red) and Band 7 (NIR) reflectance could be used as a measure of relative greenness, location-to-location, cycle-to-cycle, and location-within-cycle deviations would likely occur as a large source of error. Thus, the difference in Band 7 and Band 5 reflectance values, normalized over the sum of these values, is used as an index value and is called the Normalized Difference Vegetation Index (NDVI) [Rouse et al., 1974] [Tarpley et al., 1984] [Townshend et al., 1985] [Tucker et al., 1986], which is now used as a standard form of band ratio for vegetation studies.

$$NDVI = \frac{NIR - Red}{NIR + Red} \qquad (3.2)$$

NDVI is quite invariant against light intensity changes, it however behaves differently due to reflectance spectra changes. This explains why NDVI approaches are not really applicable under strong/low sunshine conditions where illumination effects occur.

3.1.3 Perpendicular Vegetation Index

Remarkably, the wet soil is usually confused as chlorophyll-less vegetation because the reflectance of NIR from water and soil is also very strong. In addition, the presence of soil background affects to the distribution of reflectance spectra where the expected hyperplane which aims to classify vegetation and non-vegetation in NIR-Red space is no longer passing the origin (see **Fig. 3.4**). Thus, [Richardson & C. L., 1977] suggested to measure the distance in the scatter plot from the soil line, then, pursued this approach with the Perpendicular Vegetation Index (PVI).

$$PVI = sin(\alpha)\rho_{NIR} - cos(\alpha)\rho_{Red} \qquad (3.3)$$

3.1. Related Work

Figure 3.4: Scatter plot of NIR reflectance vs. Red reflectance for all pixels in a typical image. Different regions in the scatterplot clearly correspond to different types of pixels in the image. Pixels in the green region correspond to vegetation, and pixels in the blue region correspond to sky [Bradley et al., 2007].

Where α is the angle between the soil line and the NIR axis. With an additional variable α, the hyperplane could be modified to be forward or backward the origin.

3.1.4 Difference Vegetation Index

A common special case of PVI is when α equals 45°. Here the PVI is simplified to what has been called Difference Vegetation Index [Lillesand & Kiefer, 1987].

$$DVI = \rho_{NIR} - \rho_{Red} \qquad (3.4)$$

3. VEGETATION INDICES

3.1.5 Soil-Adjusted Vegetation Index

[Huete, 1988] introduced a soil-adjusted vegetation index(SAVI) to minimize soil brightness influences from spectral vegetation indices involving Red and NIR wavelengths.

$$SAVI = \frac{NIR - Red}{NIR + Red + L} \times (L+1) \quad (3.5)$$

The constant L is added to shift the origin toward negative values to a point where intermediate densities of vegetation converge with the soil line. So, a measure of distances in the scatter plot from the soil line reveals vegetation index. However, in different lighting conditions and different circumstances, suited L values are unpredictable, which have to be manually adjusted. This degrades the applicability of this index in an autonomous process, thus, NDVI is still more preferable and well-known in this field.

3.1.6 Modified Soil Adjusted Vegetation Index

[Qi et al., 1994] provides a formula for automatically determining L from the current image data in their Modified Soil Adjusted Vegetation Index (MSAVI). The closed form of this formula is known as MSAVI2 [Jordan, 1969].

$$MSAVI2 = \frac{2(\rho_{NIR-Red}+1) - \sqrt{(2\rho_{NIR-Red}+1)^2 - 8(\rho_{NIR} - \rho_{Red})}}{2} \quad (3.6)$$

3.2 A Novel Vegetation Index : Modification of Normalized Difference Vegetation Index

The work of [Bradley et al., 2007] proposed the combination between vegetation indices and 3D-point distribution. Accordingly, the precision of vegetation detection can be reached to 95.1%. Nevertheless, such results might be obtained for scenes captured in regular environments but not clutteRed ones and also under *fine* sunshine

3.2. A Novel Vegetation Index

conditions. Under strong/weak sunshine conditions or in complex environments, the performance degrades sharply. First, with the presence of shadow, shining, underexposure or overexposure effect, the vegetation indices behave differently due to non-linear changes of NIR and Red reflectance. This is explained through the mathematics expressions of those indices in **Eq. 3.2**, **Eq. 3.3**, **Eq. 3.4**, **Eq. 3.5**, and **Eq. 3.6**, where vegetation indices are positively proportional with NIR but negatively to Red. Second, the use of laser scanner's data defeats the purpose of real-time because of time-consuming in data acquisition (2s in average to acquire a frame with 6437 scanned points).

In order to overcome the strong affection of the sunlight to the work of detecting vegetation in outdoor environments, a suggestion to use a new vision system which consists of itself near infrared lighting source has been investigated in this work. A modification of the normalized difference vegetation index is devised, which is then defined as the new standard form of vegetation index for such vision system integrated with an additional lighting system. Finally, we will show the out-performance of the proposed approach in comparison with more conventional ones.

The more details will be drawn as follow. Subsection 3.2.1 derives the novel index applied for detecting vegetation when an additional NIR lighting source used. Section 3.3 illustrates experiments and results while section 3.4 concludes this work.

3.2.1 Derivation of Novel Index

Although NDVI is well known as the standard form for vegetation index, it is no longer represent normalized difference vegetation index under a strong/low sunshine condition and with a presence of shadow, shining, overexposure or underexposure effect. Practical experiments show that the changes of NIR and Red reflectance are not linear; concretely the change of the NIR reflectance is much superior than of the Red one. Thus, if consider two parts of a vegetation region where the first part is strongly shined by the sunlight and the other part is coveRed by a shadow (see **Fig. 3.5**), the NDVI of the second part is much superior than of the first one.

Proof: Let $NDVI_1 = (NIR_1 - Red_1)/(NIR_1 + Red_1)$ and $NDVI_2 = (NIR_2 - Red_2)/(NIR_2 +$

3.2. A NOVEL VEGETATION INDEX

Figure 3.5: Illustration of variations in viewing and illumination conditions for real-world scenes containing vegetation. The vegetation varies in imaging scale and are imaged under different outdoor lighting conditions (Samples of the data can be downloaded here: http://duong-nguyen.webs.com/vegetationdetection.htm).

Red_2) represent the normalized difference vegetation indices of a vegetation in two different lighting conditions (thus, expected $NDVI_1 \approx NDVI_2$). Assume $NIR_1 \approx NIR_2$ and $Red_1 >> Red_2$, so if $\alpha = NIR_1/Red_1$ and $\beta = NIR_2/Red_2$ then $\alpha << \beta$. $NDVI_1$ and $NDVI_2$ are written as:

$$NDVI_1 = 1 - \frac{2.Red_1}{NIR_1 + Red_1} = 1 - \frac{2}{\alpha+1} \qquad (3.7)$$

3.2. A Novel Vegetation Index

$$NDVI_2 = 1 - \frac{2.Red_2}{NIR_1 + Red_2} = 1 - \frac{2}{\beta + 1} \qquad (3.8)$$

Thus $NDVI_1 \ll NDVI_2$ due to $\frac{2}{\alpha+1} \gg \frac{2}{\beta+1}$. This destroys the meaning of "Normalized Vegetation Detection Index".

The MultiCam uses an active lighting system to send modulated NIR sinals and receive reflected NIR signals through the PMD senor, so that it is not much influenced by the shining, shadow and under-overexposure effects. In fact, NIR intensity information is obtained from the modulated light while the colour information is captured from the sunlight reflection. Therefore, the standard form of evaluating radiance of light bands, or NDVI, is also not relevant for vegetation index. Practical experiments have shown the following properties of vegetation regarding light absorption/reflectance spectra of vegetation.

Property 1: The vegetation areas reflects NIR light stronger than others. In other words, NIR intensity values of the vegetation regions in the NIR image are higher than of others.

Property 2: The brighter the higher NIR intensity value is.

Property 3: Chlorophyll-rich vegetation strongly absorbs Red and blue light.

From the **Property 1** and **Property 2**, thresholding NIR values seems to be detecting bright areas. From the **Property 1** and **Property 3**, thresholding NDVI values seems to be detecting dark areas because the NDVI is negatively proportional with the Red expressed in **Eq. 3.2, Eq. 3.7, Eq. 3.8** (see **Fig. 3.6**).

Practical experiments also show that the relation between NIR, Red, NDVI and Luminance is somehow proportional but not linear, so a non-linear training technique was firstly proposed in this work. From that, we hand-labelled vegetation areas in order to extract the corresponding information (NIR, Red, Luminance) which were then gathered to form feature vectors used as training data. Support vector machine with radial basic kernel was used to train and test the results for 1000 scenes captured from both morning and afternoon conditions. The results are quite appreciated with more than 93% of accuracy, however it is time-consuming due to a large amount of points need to be trained and evaluated. Also, using machine learning technique

3.2. A NOVEL VEGETATION INDEX

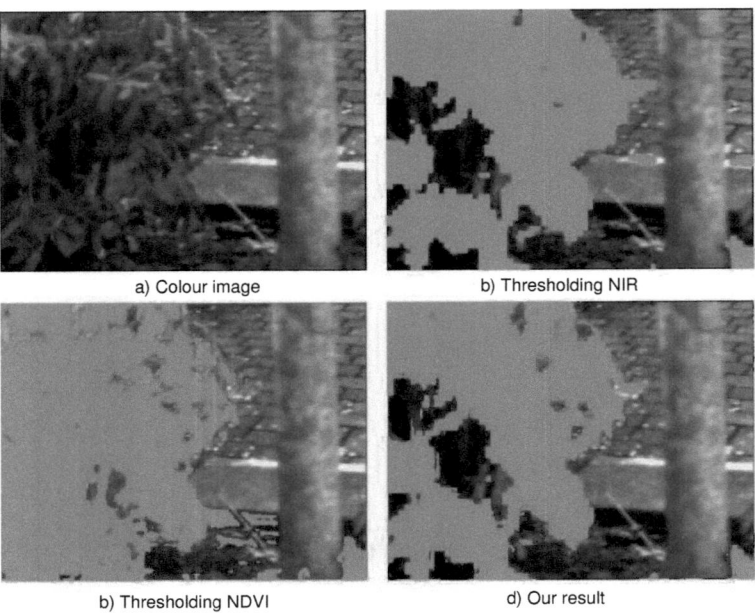

Figure 3.6: Examples of our vegetation detection result compared with thresholding NIR and NDVI.

seems to be ad-hoc where the influences of each individual term of NIR, Red and Luminance are unknown.

Alternatively, regarding the typical reflectance and absorption properties of vegetation, the impact of Luminance of the sunlight on the NIR and Red reflectance of vegetation areas is illustrated in **Fig. 3.7 (Left)** while NIR-Red wavelength space is sketched with selected vegetation points drawn as green circles in the left picture of the **Fig. 3.7 (Right)** (1.172.929 points are hand-labelled and selected in our case).

Accordingly, **Fig. 3.7 (Left)** shows an approximately linear proportion of Luminance to Red but a logarithm proportion to NIR. Again, the distribution of the vegetation points in **Fig. 3.7 (Right)** reveals that the hyperplane to classify vege-

3.2. A Novel Vegetation Index

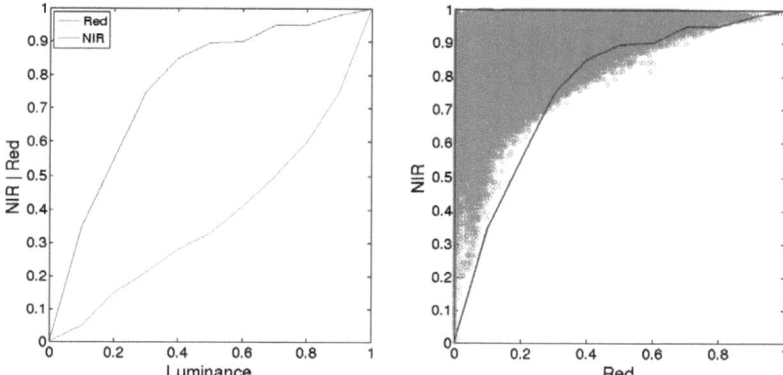

Figure 3.7: **(Left)** The impact of Luminance on NIR and Red reflectance (normalised grayscale correlation) in vegetation areas. **(Right)** Vegetation samples are sketched on the space NIR-Red as green circles, the impact of Luminance on NIR reflectance is referenced as the blue line.

tation and others could be in logarithmic form instead of the linear one as resulted from the standard form of **NDVI** [Bradley et al., 2007]. This confirms a logarithm relationship between the Red and NIR information of vegetation against illumination changes (notice: NIR used here is the active NIR at 870nm, produced by our LED lighting system integrated in the MultiCam). Therefore, we expect the hyperplane in the NIR-Red space is in a logarithm form.

Or:

$$NIR = A \times log(Red + \varepsilon) \qquad (3.9)$$

Where $\varepsilon (\geq 1)$ is a constant used to avoid a negative NIR (in our case $\varepsilon = 1$). In order to test the validity of the **Eq. 3.9**, we captured 4000 scenes and 20 videos. The performance of using the hyperplane to detect vegetation is very impressive with more than 90% of accuracy. Furthermore, the most suited hyperplanes focus on the region bounded by the green and Red lines as sketched in **Fig. 3.8**.

Whereby the higher value of A set means the chlorophyll-richer vegetation supposed to be detected. Let's take several mathematics transforms based on the **Eq.**

3.2. A NOVEL VEGETATION INDEX

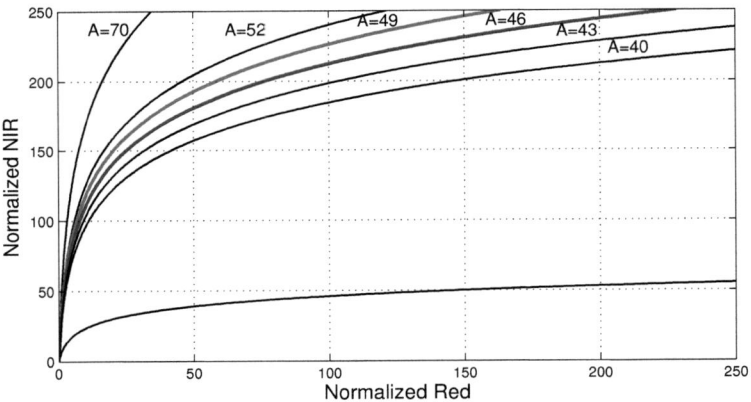

Figure 3.8: Vegetation spectra curves in NIR-Red wavelength space as predicted by the adjusted normalized difference vegetation index (in grayscale). The region bounded by the green and Red lines indicates the range of the most popular separated curves used for vegetation detection.

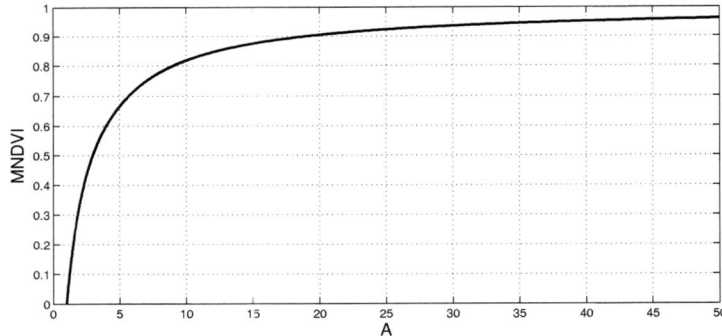

Figure 3.9: Positive relationship between the Modification of Normalized Difference Vegetation Index and the factor A.

3.9:

$$A = \frac{NIR}{log(Red+\varepsilon)} \rightarrow \frac{A-1}{A+1} = \frac{NIR-log(Red+\varepsilon)}{NIR+log(Red+\varepsilon)} \quad (3.10)$$

3. Vegetation Indices

If we denote MNDVI = $\frac{A-1}{A+1}$, so:

$$MNDVI = \frac{NIR - log(Red + \varepsilon)}{NIR + log(Red + \varepsilon)} \qquad (3.11)$$

MNDVI: so called Modification of Normalized Difference Vegetation Index.

This index ranges from 0 to 1 ($A \gg 1$, see **Fig. 3.8**), thus avoiding negative values (unlike the NDVI). The relation between MNDVI and A is sketched as in **Fig. 3.9**. Accordingly, they have a positive relation, so the MNDVI also represents the vegetation index where *the higher value of MNDVI is, the chlorophyll-richer vegetation is supposed to be detected*. Look at NDVI in **Eq. 3.2** and MNDVI in **Eq. 3.11**, they share a similar mathematics form and all express the Normalized Difference Vegetation Detection. The logarithmic term in the later formula expresses the less impact of the Red when an artificial lighting system is used.

3.3 Experiments and Results

In order to evaluate the performance of the proposed approach in comparison with previous ones, our autonomous ground vehicle took 5000 raw images and 20 videos of outdoor scenes containing vegetation, under both morning and afternoon conditions as well as shadow, shining and underexposed effects taken into account. The configuration of the AGV can be seen in [Nguyen *et al.*, 2011b], subection 2.1 where the LiDAR, CMOS camera, and MultiCam mounted at front up of the robot help in describing complex environments.

When the AGV traverses the environment, all data is collected and stored in its computer. In the first experiment, we implemented vegetation detection algorithms based on "Local point statistic" [Lalonde *et al.*, 2006], "Conditional local point statistic" [Nguyen *et al.*, 2010b] and "2D-3D feature fusion" [Nguyen *et al.*, 2011b], using CMOS and laser data. In the second experiment, we built vegetation detection algorithms found on Normalized Difference Vegetation Index using MultiCam's data. In the later experiment, the performances of vegetation detection based on Support Vector Machine training and Modification of Normalized Differ-

3. VEGETATION INDICES

ence Vegetation Index are shown in the confusion matrix as in **Table 3.1**.

Table 3.1: Confusion Matrices for Different Methods(%)

	SVM [Nguyen et al., 2011c]		MNDVI (proposed)	
	Vegetation	Others	Vegetation	Others
Vegetation	93.14	5.69	91.02	11.77
Others	6.86	94.31	8.98	88.23

Figure 3.10: The first row illustrates original colour images. The second row shows the results given by NDVI approach. The third row demonstrates the results given by the proposed approach.

The comparison between different vegetation detection approaches' performances is illustrated in **Fig. 3.10** and **Table 3.2**. **Fig. 3.10** demonstrates the out-performance of the proposed approach against NDVI approaches. Intuitively, MNDVI approach performs more robust and stable vegetation detection under different illumination effects such as shadow, shining, under- and over-exposure. In **Table 3.2**, the evaluation describes the confusion matrices of available approaches whose levels of consideration in environmental and illumination complexities and real-time constraint are relatively pointed out to assess the reliability and accuracy of the approaches. In this paper we consider five levels of illumination complexity, including: intensity-colour change, shadow, shining, underexposure, overexposure. The number of illumination effects taken into account of a approach reveals the level of illumination complexity

3. Vegetation Indices

Table 3.2: Evaluation of Vegetation Detection performances against environmental complexities (EC), illumination complexities (IC), and real-time constraint

Authors	Methods	Sensors	Constraints		Real-Time	Confusion Matrix(%)	
			EC	IC	(fps)	True Positive	True Negative
Lalonde,2006	(LPSA)	Laser	4	5	< 1	0.48	0.41
Nguyen,2010b	(CLPSA)	Laser	4	5	< 1	0.58	0.47
Nguyen,2011b	(2D3DFF)	Laser+ mono	4	3	< 1	0.84	0.69
Wurm,2009	Laser Remission	Laser	1	5	1 *up to* 3	0.99	0.93
Bradley,2007	NDVI	Laser+ (MSC)	4	4	1 *up to* 3	0.95	0.81
Nguyen,2011c	SVM	MultiCam	5	4	3 *up to* 5	0.93	0.83
Proposed	MNDVI	MultiCam	5	5	10 *up to* 14	0.91	0.85

(*) Abriviation: (LPSA) Local Point Statistic Analysis; (CLPSA) Conditional Local Point Statistic Analysis; (2D3DFF) 2D-3D Feature Fusion; (MSC) Multi-Spectral Cameras.
(**) Levels of consideration: **5:**very high; **4:** high; **3:** neutral; **2:** low; **1:** very low.
(***)The frame rate was estimated using laser scanner-LMS221.

of the approach. Alternatively, regarding complex environments, we divide environments into five levels: level 1: hall-way/in yard/campus; level 2: rough/soil road; level 3: off-road with low-grasses; level 4: off-road with tall-grass/ bushes, level 5: forest. Accordingly, the "Local point statistic" is not affected by illumination changes, and can be applied for complex environments if time is not a criteria. The figure is quite similar for the "Conditional local point statistic" but the precision is improved at about 10%. "2D3D feature fusion" provides a significant improvement of robustness where the precision reaches to 84%, it however is still a time consuming approach. To approach more the real-time constraint, "Laser remission" restricts the complexity of environments with only two classes analysed, as a result, the precision obtained is quite high, at more than 99%, the processing time is neutral (around a half second using LMS221).

Emphasizing on the task of detecting vegetation, the approaches based on the photosynthesis-related properties of vegetation enable faster and higher-precision

3. VEGETATION INDICES

processes for a real-time and robust vegetation detection system. Indeed, the combination between Normalized Difference Vegetation Index and three dimensional distribution can boot the precision up to 95% while the processing time does cost very expensive. When the purpose is really to extend the reliability and accuracy of vegetation detection in outdoor complex environments, the shadow, shining and under-exposed have to be taken into account so that a new vision set-up with an active lighting system is recommended. Actually, the performance of using MNDVI for MultiCam's data to detect vegetation is very impressive where all constraints are highly considered. In a traditional NDVI approach [Bradley et al., 2007], vehicles painted with pigments that are reflective in NIR can also be misclassified as vegetation. Meanwhile, the proposed approach uses an active lighting system which can avoid this failure in most cases. Indeed, vehicles are usually designed with smooth surfaces which reflect NIR rays from the active lighting system to another direction, thus, PMD sensor would not receive any of that reflected NIR rays. As a result, the NIR information received by PMD sensor are the NIR light reflected from the sunlight which is rather weak, thus, does not cause hue impact as in the traditional way. Still, human wearing dark clothes can be misclassified as vegetation in the both approaches due to a high infrared radiation emitted from the human body.

3.4 Conclusion

We have introduced the overview of vegetation detection in a structured way with respect to vegetation index-based approaches as well as presented our new vision set-up to completely realize the work under the real-time and robust constraints. Overall, our approach shows out-performance compared with others when taking all environmental and illumination complexities as well as real-time constraint into account. Regarding the performance of the MultiCam, the range measurement is still poor in outdoor environments, thus, the proposed approach could not use depth information for any detection application but just for obstacle avoidance. Alternatively, the wavelength of the modulated light in the MultiCam's lighting system strongly

3. Vegetation Indices

focuses on the band around 870 nm while the expected band starts from 800 to 1400 nm, so the chlorophyll less-vegetation like brown/Red/yellow grass is not well detected. However, if extending the spectral width of the modulated light, it degrades the range measurement of the MultiCam. Therefore, a compromise between range measurement and vegetation detection will be considered in our future works. An idea to produce a similar device only for vegetation detection with full band of 800 nm 1400 nm for the desired lighting system will also be taken into account for a further development of the vegetation detection system for outdoor automobile guidance.

3. VEGETATION INDICES

Chapter 4

2D-3D Feature Fusion-based Vegetation Detection

Vegetation detection is very simple for the human eye based on its typical colours, textures and geometric distributions. Thus, the idea of capturing those characteristics of vegetation intuitively has been investigated by using different visual sensors and techniques. In general, most of available approaches base on image processing (2D sensor) or point cloud analysis (LiDAR) separately. The image processing-based vegetation detection exploits colour and texture features of vegetation while the point cloud analysis-based vegetation detection examines its 3D structures. The performance of the first approach depends significantly on the illuminating conditions where images are captured. The second one fails to deal with complex outdoor environments, especially with the presence of dense edges. Both approaches could not lead to detect the variety of vegetation in nature; see more details in section 4.1. We, hence, propose a 2D/3D combination approach which can utilize the complement of three-dimensional point distribution and colour descriptor. First, a 2D/3D mapping needs to be carried out in order to obtain the correspondences between the image plane and the LiDAR 3D coordinate; see section 4.2. Second, 3D point cloud is segmented into regions of homogeneous distance, and then 3D features are extracted by implementing conditional local point statistic analysis on each region, described in section 4.3. The regions of interest segmented from the point cloud are

4. 2D-3D FEATURE FUSION

projected into the image plane to result the corresponding regions of interest. Finally, colour descriptors are studied and applied to those regions to extract colour features, see section 4.4. Those all scatter and colour features will be trained to Support Vector Machine to generate a vegetation classifier. Finally, we will show the superior performance of this approach in comparison with more conventional ones, as in section 4.6.

This work has been published in **Proceedings of IEEE ICIT-2011** [Nguyen et al., 2011b].

4.1 Related Work

As mentioned above, vegetation detection is very simple for the human eye, it however is absolutely not trivial for the robot's eye. Human eye is able to recognize reflectance changes without considering shadows and unexposed effects; contrariwise, using image processing techniques, an increasing or decreasing in reflectance could happen under different lighting conditions. Indeed, regarding the view-point of image processing, first there is no specific shape and texture of general vegetation. Second, although vegetation normally owns typical colours such as green, red orange, and yellow, the colour descriptor-based vegetation detection is unstable due to light colour and light intensity changes under different sunshine conditions in outdoor environments. So, it should be made clear that many publications regarding pattern recognition mentioning grass/leaf detection successfully by using texture and colour information [Zafarifar & de With, 2008][Lu et al., 2009][Wu et al., 2004][Manduchi, 1999], they however indicated some specific species of vegetation but not vegetation in general. As a consequence, those approaches were just applied for robots operating in structured environments but not cluttered ones as investigated in this work. Overall, the only use of colour and texture information cannot result a robust vegetation detection in complex outdoor environments, which leads researchers to come up with the other distinct features rather than colour descriptors, or combine many of them.

Regarding the literature of robotics research, vegetation, especially grass, is detected as one class in several classes of classified terrains used for determining navigable or non-navigable terrains [Wolf & Fox, 2005][Dahlkamp, 2006][Rasmussen, 2001][Manduchi, 2005]. Those approaches model ground surface and objects above the ground are generally obstacles. [Wellington et al., 2006] introduced a more advance approach which models more complex terrains to learn which obstacles can be driven over (low grass, ground) and which need to be avoided (bushes). Indeed, [Wellington et al., 2006] modelled terrain structure as a set of voxels where each voxel is a $15cm^3$ box-shape region of three dimensional space. The simple idea of detecting vegetation can be explained as follows: the number of LiDAR rays that pass through each voxel is recorded (pass-through); the number of LiDAR rays that hit that voxel is also recorded (hits); thus, the voxels which contain mixture of hits and pass-through should be vegetation. To improve the robustness, the work introduced Markov Random Field (MRF) models and Hidden semi-Markov models (HSMM) to model 3D structure of terrain based on laser remission, infrared temperature and colour information. However the approach requires some constraints which help for a better navigation but limit the applicability of the method for detecting a variety of vegetation. For instance, the state between the ground height and vegetation height is vegetation and above that is free- space; the similarity in vegetation height; etc. In this work, we are more interested in approaches which can be applied for detecting a variety of vegetation in nature.

4.2 2D/3D Mapping

The problem of calibrating a vision system is extremely important for practical applications such as 3D reconstruction and pose estimation of three-dimensional objects. Even though many researchers attempted to do full-calibration of coupled vision systems such as Fish-eye Laser Scanner and CCD camera or CMOS camera, the result showed mean performance while the cost of computation was very expensive [Brun & Goulette, 2007]. The precision of reconstructing 3D model drops sharply with the

4. 2D-3D FEATURE FUSION

presence of vegetation. One of the main reasons is that interest points are not stable due to the vibration of vegetation. In fact, for the aim of detecting vegetation, we do not need a very precise calibration. A simple 2D/3D mapping with all large objects reconstructed is sufficient. Therefore, we on the other hand propose a simple but fast and efficient 2D/3D mapping technique for the coupled system: Laser Scanner and CMOS camera. The characteristics of elements are described in the **Table. 4.1**.

Table 4.1: Characteristics of Elements

Laser Scanner	Number of points	6437 per profile
	Aperture angle	$41^0 \times 77^0$
	Profiles velocity	2 s
	Focal length	12 mm
Colour CMOS	Number of points	640x480 pixels
	CMOS	$\approx 1/3.2$"
	Frame rate	25 Hz
	Aperture angle	$55^0 \times 70^0$
	Focal length	3.7 mm

The technique is found on the following property:

Property 1: *If the CMOS camera and Laser Scanner are positioned near each other in a vertical line, and when objects are far enough, the views from CMOS camera and from Laser Scanner are nearly the same in a narrow angle. So, a simple 2D/3D coarse calibration can be done by mapping two images lied on two parallel coordinates.*

Hence, we need two assumptions: At first, CMOS camera and Laser Scanner have to be positioned near each other and in a vertical line. Secondly, all object are far enough to avoid the stereo effect. One might concern wheather these assumptions are strong or weak ? In order to answer the question, we have positioned CMOS camera as 5 cm under LMS221, and tested for the performance of the coupled system, see **Fig. 2.2** in section 2.1. The views of LMS221 and CMOS camera are quite correlative for the object distance of 3.8 m or more (see **Fig. 4.2 a) b) c)**). Specifically, the practice, with 525 scenes captured, has proved that the assumptions are strong for purpose of vegetation detection in the distance range of 3.8 m to 15.8 m. The

4. 2D-3D Feature Fusion

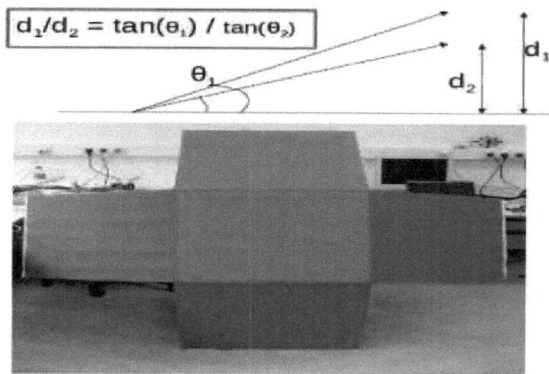

Figure 4.1: The proportion of size of CMOS image to depth image's is equal to the proportion of aperture of CMOS to LMS221's, in each dimension. The 3D model is created by Johannes Leidheiser, Lars Kuhnert and Klaus-Dieter Kuhnert, see more in Leidheiser [2009].

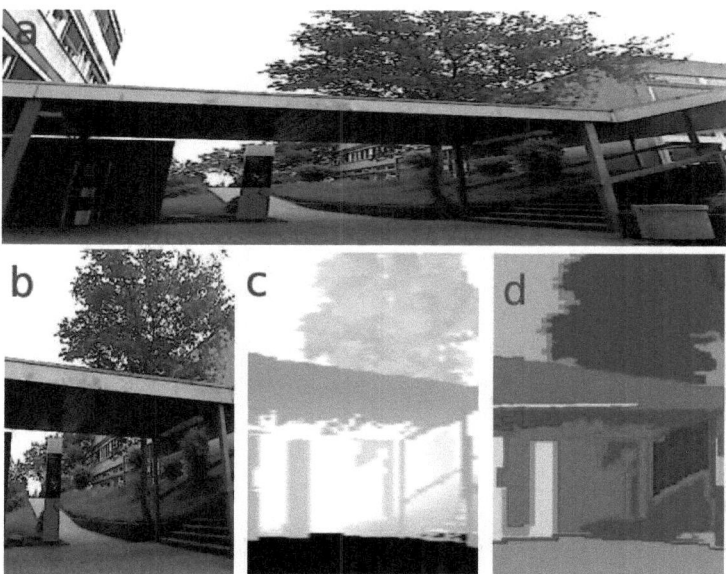

Figure 4.2: a) CMOS image, b) cropped CMOS image c) depth image d) segmented image.

4. 2D-3D FEATURE FUSION

upper threshold for distances needs to be set due to the scatter effect of laser beam, whereby very far objects will not be captured enough information to be recognized regarding 3D point distribution. To calibrate the coupled systems, a 3D chessboard model is built, see **Fig. 4.1**. The procedure of 2D/3D mapping is taken place as following.

1. Finding the size of the depth image projected to the CMOS image, denoted by L_{Size}(width, height). The proportion of size of CMOS image to depth image's is equal to the proportion of aperture of CMOS to LMS221's, in each dimension (see **Fig. 4.1**).

2. The depth image is interpolated to the size of L_{Size}. The technique used is linear interpolation.

Figure 4.3: Example of reconstructed 3D scenes.

4. 2D/3D Feature Fusion

3. Sliding a window with size of L_{Size} across the CMOS image and do matching with the interpolated image. Four interesting points matched are the conner points of the models. Considering the centroid of each image, CMOS image or depth image, as the origin of the corresponding image plane: Assume that $O_1(0,0)$ is the centroid of the CMOS image, and $O_2(x_{shift}, y_{shift})$ is the centroid of the depth image projected on the CMOS image plane. The matching process returns the shifting parameters (x_{shift} and y_{shift}) of the two image planes. The unit of these parameters is pixel size with 640 x 480 of CMOS image resolution.

After the mapping, we will obtain three parameters including L_{Size}, x_{shift} and y_{shift}. These parameters are then used to reconstruct 3D scene as seen in **Fig. 4.3** where L_{Size} = 398x472 pixels, $x_{shift} = 19$ pixels and $y_{shift} = -74$ pixels.

4.3 3D point cloud analysis

The traditional way for analysing 3D data given by a LiDAR is to capture the spatial distribution of points in local neighbourhood [Vandapel et al., 2004]. In this work, we, on the other hand, address a statistic approach based on 3D point distribution for analysing the 3D data. At the early state of our work, we have done the local statistic analysis for 3D point cloud given by the LMS221 (LiDAR)[Nguyen et al., 2010b]. The basic idea of this approach is that the point clouds representing artificial constructions and tree trunks should have linear or surface structure while the vegetation should be represented by high textured or scattered structure clouds. The work of [Lalonde et al., 2006] demonstrated that this idea can be potentially implemented in describing outdoor environments. However, the task of finding an efficient way to classify the 3D structures is very challenging.

The first suggestion is introduced by [Lalonde et al., 2006]. A sizeable cube crosses by the 3D point cloud to capture the local spatial point distribution by the decomposition into principal components (PCA) of the covariance matrix of the 3D point positions, ordered by decreasing eigenvalues. Intuitively, in the case of scattered points, there is no dominant direction in the spatial distribution of the points,

4. 2D/3D FEATURE FUSION

so the eigenvalues are nearly equal to each other. In the case of linear structure, there should be only one dominant direction, so the first eigenvalue is much superior to the others. Finally, in the case of solid surface, the principle direction is aligned with the surface normal with the first two eigenvalues are close to each other and far different from the last one. These properties are very efficient and characteristic to describe the 3D spatial distribution of points in space. However, the way of sliding a cube across the space to estimate the local point spatial distribution is time-consuming and can not deal with the variety of environments. The suitable sizes of the cube are unpredictable and must be adjusted in different conditions. Besides, in the case of dense edge presence in artificial constructions, a set of edge points will look like a porous volume which defines the character of vegetation. Therefore, we have proposed a pre-processing procedure that segments 3D point cloud into regions of homogeneous distances. The segmentation will help to avoid the edge effect ([Nguyen et al., 2010b]) and extract objects in forms of region interest. As also discussed in our previous work that the segmentation technique should be applied is Efficient Graph-based [Felzenszwalb & Huttenlocher, 2004]. The technique covers both local and global properties of images, which is proved neither too coarse nor too fine in the work of [Felzenszwalb & Huttenlocher, 2004]. Indeed, the distances given by the LMS221 are very precise. So, the image representing the point cloud are more structured and finer than a regular image whereby the segmentation algorithm has no longer to face illumination effects of natural scenes. Indeed, the segmentation of 3D point cloud has been carried out successful in our previous work [Nguyen et al., 2010b], but for the purpose of mapping 2D/3D, it is time-consuming. Because, we after segmented the point cloud, then had to project it back into the CMOS image plane, and do matching with the CMOS image. In this paper, we are going to directly segment the depth image given by a projection of 3D point cloud into the CMOS image plane. The segmentation takes 42 ms for such image sizes of around 41x157 pixels. The details of Efficient Graph-based technique used for image segmentation is reference in [Felzenszwalb & Huttenlocher, 2004], while one result example is illustrated in **Fig. 4.2 d**.

4. 2D-3D Feature Fusion

4.3.1 Scatter Feature Extraction

A. Conditional Local Point Statistic

In the work [Nguyen *et al.*, 2010b], we have proved that three saliency features, including $S_{scatter}$, $S_{surface}$, and S_{linear} are efficiently used to classify terrain. Whereby we are able to classify scatter (tall grass, canopy, needle tree, thin bushes) from surface (wall/flat road/building), and linear (wire) objects. The details of how to extract those conditional local point statistic features are described in [Nguyen *et al.*, 2010b] or in subsection 8.2.2.2 in Chapter 8. However, there exists some kind of vegetation such as thick bushes, low grass, and broad leaves trees which could not be distinguishable from surface objects when applying the local point statistic technique. Also, in complex environment with presence of dense edge objects, the judgement of scatter objects being vegetation is not always applicable. Therefore, a clever strategy is to remove all smooth surface and linear objects out of interest by applying machine learning techniques on the set of these conditional local point statistic features. Rough surface objects and objects with porous volume in 3D structure are more investigated with their level of roughness and colour features.

B. Regional Distance Distribution

Regarding the representation of distances from the point of view of image processing, a histogram of depth images can describe the distribution of distances. In this work, we will show that the histogram of distances can be used efficiently to estimate the scatter property of points in space. We have done many experiments which prove that 20 bins of histogram is sufficient for the estimation. A histogram model H_s is assumed to be built beforehand by averaging histograms of samples with scatter distribution.

$$H_s[i] = \frac{1}{N} \sum_{j=1}^{N} H_j[i] \qquad (4.1)$$

where N: number of scatter samples, $H_j[i]$ is the value of the bin i in the jth histogram. For each region R_k, distances firstly are normalized into the range of [1 20]. The histogram of each region is H_k which has to be normalized with H_s,

4. 2D-3D FEATURE FUSION

or: $\sum_{i=1}^{20} H_k[i] = \sum_{i=1}^{20} H_s[i]$. The quadratic histogram distance between H_k and H_s is computed as following.

$$D_k = sqrt((H_k - H_s)^T * A * (H_k - H_s)) \qquad (4.2)$$

where $*$: denotes the convolution. Matrix A describes the internal element difference of H_s.

$$A_{ij} = 1 - \frac{|H_s[i] - H_s[j]|}{argMax_{i,j}(H_s[i] - H_s[j])} \qquad (4.3)$$

$(H_k - H_s)^T * A * (H_k - H_s)$ is positive semi-definite on the subspace $\sum_i (H_k - H_s)[i] = 0$, so $D_k^2 \geq 0$. The distance D_k is then used as a scatter feature. Even though, the histogram quadratic distance is computationally expensive for a big number of bins, it is very efficient and fast for such 20 bins in this work.

4.4 Colour Descriptors

The human eyes' perception of colour is one of the most important visual elements which help us to recognize different objects. In addition, vegetation does not have a specific shape or texture but usually represented by green, orange, or yellow colour. Therefore, this work pays more attention on studying colour descriptor. In fact, the work of [van de Sande et al., 2010] has introduced a very structured overview of different colour invariant descriptors in the context of image category recognition. The colour invariant descriptors have been evaluated individually where high precision of detecting specific objects such as aeroplane, person, horse, and car is shown. However, the detection of vegetation, in particular potted plant, is still very poor, at about 20% in average precision. One of the major problems for that is the shift and change of intensity and colour under different light conditions while the vegetation tends to be recognized based mostly on its colour. So, two of interesting features should be taken into account are the mean and standard variation values of intensity and colour which imply the light condition of the viewed scene. The interesting point in veg-

4. 2D-3D Feature Fusion

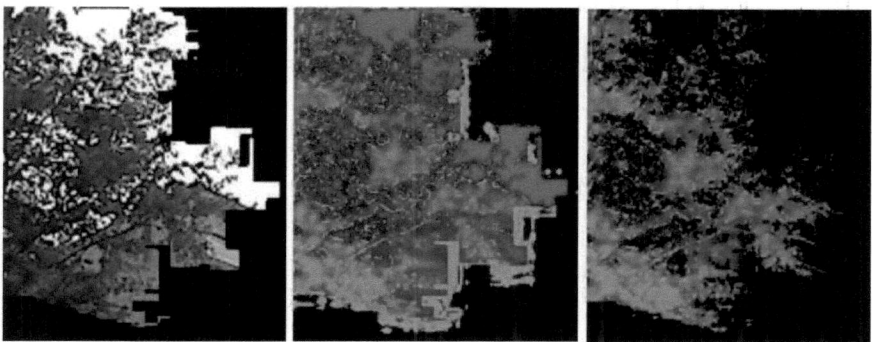

Figure 4.4: a) an example of vegetation regions extracted from the section III. b)Raw hsv image c) hsv image after thresholding Value's intensities.

etation images is that the main colour should be theoretically green in HSV colour space under most different environment conditions. In reality, this is not always true for scenes containing sky. The affection of sky tends to turn the colour of image to red, red brown, etc. The issue is often caused by the low intensity of the "Value" (in HSV colour space). Therefore, this can be solved by giving the lower threshold for the Value's intensities: If $V[i] < \kappa$ then $HSV[i] = 0$.

The result is illustrated in **Fig. 4.4** (It should be implicitly understood that an image pre-processing step needs to be done beforehand such as noise and blur filtering). The green or orange colour appears as a majority colour in vegetation images in HSV colour space. This drives us to come up with a vegetation recognition based on colour histogram distribution. That means the global properties of colour in an image are more emphasized than the local ones.

Colour histogram distribution is not new in content based image retrieval (CBIR) and image category recognition [Hafner et al., 1995][Jeong et al., 2004][van de Sande et al., 2010]. However, the use of it in detecting vegetation or categorizing vegetation images with others has not been done successfully up to now. In order to focus more on vegetation detection, we propose a histogram model, denoted by H_v, which is obtained by averaging histograms of vegetation images, or: $H_v[i] = \frac{1}{N}\sum_k H_k[i]$. N is the number of vegetation images inputted. Two common

4. 2D-3D FEATURE FUSION

histogram distances often used to compare histograms are Histogram Euclidean and Histogram Intersection are also studied in this paper. The difference of histogram H_k and H_v can be calculated in different distance definitions as following.

Euclidean Distance:

$$d_e = sqrt(\sum_i (H_k[i] - H_v[i])^2) \qquad (4.4)$$

Histogram Intersection

$$d_i = \frac{\sum_i min(H_k[i], H_v[i])}{\sum_j H_v[j]} \qquad (4.5)$$

The work of [Jeong et al., 2004] has showed the out-performance of Histogram Intersection in HSV colour space compared with it in RGB colour space and with Euclidean Distance in both colour spaces. However, the result is just applied for standard images where good light condition is set. In worse light conditions, there is a shift and change of colour. The consideration of absolute value difference of histogram bins is no longer appropriate. So, the result shows poor performance, at about 30% → 0.4% in average precision for both the above distance measurement methods. The interesting point is that there is similarity between histogram curves of vegetation images under different environmental conditions. Therefore, we suggest to use Histogram Quadratic to measure the similarity and dissimilarity between two histogram curves.

Histogram Quadratic:

$$d_q = sqrt(\frac{1}{M}(H_k - H_v)^T * A * (H_k - H_v)) \qquad (4.6)$$

$$A_{ij} = \frac{|H_v[i] - H_v[j]|}{max_{m,n}(H_v[m] - H_v[n])} \qquad (4.7)$$

where: M is the number of histogram bins; H_k has to be normalized $\sum_i H_k[i] = \sum_i H_v[i]$; A is the cross correlation matrix of histogram bins of H_v. So, A can be computed beforehand to reduce the on-line computation.

4. 2D-3D Feature Fusion

The Histogram Quadratic is firstly introduced by [Hafner et al., 1995] for image retrieval based on colour histogram. However, it seems to be computationally very expensive because the number of histogram bins is usually large. In fact, in HSV colour space, there is mostly no identity of vegetation expressed through histograms of Value and Saturation. The Hue histogram describes more characteristics of the viewed scene. This enables a quantization of the histograms in order to reduce the number of histogram bins or makes Histogram Quadratic more applicable. Even though the histogram curves share some similarities regarding global shape but they are not respectively correspondent with their histogram bin mapping values. So, the increase of quantization levels does not accompany with the increase of accuracy of histogram similarity measurement. A large number of tests has been carried out in our laboratory to choose the best quantization levels of different colour channels in HSV colour space. Practical shows that the proportion **20:4:3** corresponding to Hue:Saturation:Value results highest accuracy in classifying vegetation's histogram curves and others'. The evaluation of vegetation retrieval is illustrated in **Fig. 4.5**. Two metrics for retrieval effectiveness are *recall* and *precision*. Recall signifies the vegetation images in the database which are retrieved. Precision is the proportion of the retrieved images that are vegetation. Let A be the set of relevant items, B the set of retrieved items.

Thus: $recall = P(B|A)$ $\qquad\qquad Precision = P(A|B)$

4.5 Support Vector Machine

It is practically infeasible to hand-tune thresholds to use directly the saliency features to perform classification because those feature values may vary depending on the type of environments, the type of sensors, the number of scanned points, and the point density. Experimental research shows that the variability of the values is manifested especially with the presence of tall grass (or dense edge areas), so we usually confront with a nonlinear classification problem in the case of cluttered environments. In fact, there are many nonlinear classification techniques proposed at

4. 2D-3D FEATURE FUSION

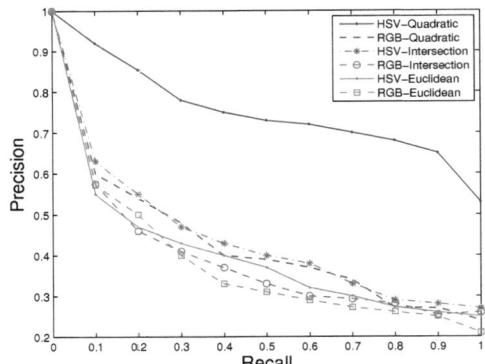

Figure 4.5: Histogram-based retrieval effectiveness for vegetation.

recent time, both supervised and unsupervised. While the supervised techniques usually cost computational expense, the unsupervised ones are not well adapted to the nonlinear problems in reality. In this work, we train data with only six features (see **Table 4.2**) in order to give the decision, so the supervised classification technique is preferred.

Table 4.2: Six extracted features

Intensity	Colour	3D distribution
Mean, Std	d_q	$S_{scatter}, D_k, S_{surface}$

We had tried to use Support Vector Machine (SVM) [Cortes & Vapnik, 1995], Naive Bayes classifier [Quinlan, 1993], Neuron Networks [Zhang, 2000], Adaboost [Freund & Schpire, 1997], and Expectation Maximization [Bilmes, 1997]. Consequently, proposed by Cortes and Vapnik in 1995, SVM shows out performance and is more reliable than others. The kernel trick used is Radial Basic Kernel [Baudat & Anouar, 2001].

4.6 Experiments and Results

In this work, 500 different scenes of cluttered outdoor environments are captured by the SICK laser LMS221 with 41x157 pixels resolution and the Logitech QuickCam Pro 9000 with 640x480 pixels resolution, in both morning and afternoon conditions. The maximum distance set is 16 m. 300 pairs of 3-D point clouds and CMOS images are used for training and the other 200 are used for testing. The classification results are evaluated by comparing the output of the classifier with the hand-labeled data. **Table. 4.3** shows the classification accuracy of the results.

Table 4.3: Confusion Matrix (%)

	Vegetation	Others
Vegetation	81.48	18.52
Others	15.76	84.24

The classification processing time of our approach is fast, at around 580 ms, however the acquisition time of the LiDAR is quite slow, at around 2000 ms. Therefore, the total processing time of this approach is at around 2580 ms, which is not really reliable for on-board navigation. The main use of this approach is to predict the scene category front of the vehicle and interpret the current environment by localizing vegetation areas around. In reality, outdoor autonomous navigation has to face with unknown environments and unknown situations. Whenever, the automobile gets into

Figure 4.6: Some vegetation detection results obtained from the proposed method.

4. 2D-3D FEATURE FUSION

a tough situation where he could not find which way to go or all ways seem to be blocked by lethal obstacles. In such situation, the approach initiates a solution. The robust detection of the proposed method enables a more interaction between automobile and natural environments. The knowledge about location of vegetation areas around gives information not only referenced for navigation by making decision of which obstacles can or can not be driven, but also for localization. **Fig. 4.6** illustrates some vegetation detection results obtained from the given method. In general, vegetation is well detected, except the case that vegetation is very far from the robot so that there is no 3D information about it, or the background light is too dark or too bright where the colour information is too bad to extract 2D features of vegetation.

4.7 Conclusion

We have presented a new approach for vegetation detection, which is based on 2D/3D feature fusion. The 3D point distribution has been investigated and fused with the colour descriptors to form the feature vectors of vegetation, which are then fed to the training process using support vector machine to generate vegetation classifier. Through many robotics experiments in a large variety of object scenarios, the classifier shows impressive performance with accuracy of more than 82%. Thus, the proposed method can be used to support decision-making of a robot in its autonomous navigation or other agricultural applications which require to detect vegetation. The limitation of the approach is that it requires a fully scanned 3D scene while all available Laser Scanners need a long data acquisition process in order to result a reasonable point cloud resolution. This lowers the speed of the entire sequence. The outlook of this work is to develop a full functional system whose precision and time can be compromisable by reducing or increasing the number of scanned points. Whereby, the system can be applied in different levels of precision and processing time depending on purpose of a given task.

Chapter 5

General Vegetation Detection Using an Integrated Vision System

In this paper, we address a new vegetation detection model by using an integrated camera mounted both CMOS and PMD sensors, the so called MultiCam, which can simultaneously provide NIR, colour, and depth images. The aim is to deal with either obstacles avoidance or scene category prediction in order to give a better decision-making framework for autonomous navigation while still considering real-time constraints. The novelties of this work lie in the following aspects:

- Chlorophyll of vegetation significantly absorbs visible light, especially red and blue wavelengths while strongly reflects NIR bands. Thus, the ratio of radiances in the NIR and red bands leads to index vegetation.

- Observing that the strong reflectance of vegetation in the modulated NIR band helps to acquire fine depth information for vegetation areas in the scene while the depths of other objects usually consist of much noise (well-known issue of PMD sensor in outdoor operating conditions). This enables good spatial features for vegetation detection based on both estimating the probability

4. 2D-3D FEATURE FUSION

of depth measurement noise and analysing the 3D distribution of vegetation point-cloud.

- Vegetation usually appears in several typical colours such as red-orange, yellow, and green. So, if we build three colour histogram models corresponding with the three colours, histogram distances between an input and the models can be used as good colour features to detect such homogeneous colour object like vegetation.

- The texture of vegetation is generally complex and unstructured, thus, an assessment on texture orientation in a local region tends to provide discriminative texture features.

Whereby, the reflectances of the modulated NIR given by the PMD sensor and the red channel of the CMOS sensor are used to calculate Normalized Difference Vegetation Index (NDVI). In addition, a relative distance estimation method referencing a perfect flat ground is described to obtain quickly 3D point cloud in the vehicle frame, thus, enables a 3D distribution analysis for spatial feature extraction. Since vegetation usually appears in several typical colours and unstructured texture, the paper also derives a method for generating colour histogram models and assessing unstructured texture orientation to create visual features. Finally, NDVI, spatial features and visual features are gathered to form feature vectors, which are then used to train a robust vegetation classifier. In [Nguyen et al., 2011c], we only used the colour and NDVI features together with a training method to create vegetation classifier. We show through empirical results that the vegetation detection accuracy is improved by taking texture and spatial features into account. In all real world experiments we carried out, our approach yields a detection accuracy of over 90%.

This paper is structured as follows. Subsection 5.1 presents the setup of our vegetation detection system. Subsection 5.2 introduces spatial features extracted from MultiCam data. Subsection 5.3 mentions the fast calculation for normalized difference vegetation index. Subsection 5.4 discusses on colour and texture features of vegetation from human perspective. Subsection 5.5 demonstrates some experiments

5. General Vegetation Detection Using an Integrated Vision System

and results while Subsection 5.6 concludes this work.

A part of this work has been published in **Proceedings of IASTED International Conference on Computer Graphics and Image Modeling** [Nguyen *et al.*, 2011c]. This work has been accepted for publication in **International Journal of Robotics and Automation** [Nguyen *et al.*, 2013].

5.1 System Set-Up

Due to the high frame rate of data acquisition, MultiCam is well suited for a real-time object detection whereby interactive 3D data can be obtained by moving the camera around in space. Indeed, many applications have been implemented successfully on using a PMD camera such as hand detection, man-machine interaction, and obstacle detection [Ghobadi *et al.*, 2010]. These firstly motivated us to extend the use of the camera in a diversity of other applications confidently. However, a huge problem is arising regarding to the high probability of failed distance measurement whenever the camera is working under strong sunshine conditions. There are actually many aspects affecting the degradation of measure, which were mentioned in [Nguyen *et al.*, 2010a] in details. Typically, we just want to emphasize: the illumination noise caused by abundant NIR of the sunlight; the significant influence of the distance aliasing effect [1] of measured distances relied on a phase modulation process used in ToF cameras. In case of the alias occurring to some foreground objects and background, the measured distance results contain no benefit information for those objects. Actually, this is not a big deal or not a frequent occurrence in indoor structured environments such as hall-way and room space. In contrast, the outdoor scenes include so many different objects at different distances, even the sky, which inevitably challenges the distance aliasing effect. In fact, the purpose of terrain classification as well as vegetation detection in order to help an AGV operate outside does not require all knowledge of surroundings but the understanding of the front

---

[1] The distance aliasing effect is understood in this case as the overlap of measured distance values for objects whose distance differences between them are multiples of 7.5m. For example, a target at 8.5 m would appear as at the distance of 1 m (= 8.5 m - 7.5 m) from measuring (understood as a circular shift by the phase modulation process used in the ToF camera).

5. GENERAL VEGETATION DETECTION USING AN INTEGRATED VISION SYSTEM

scene. Therefore, we propose to position the MultiCam looking down as can be seen in **Fig. 5.1(b)**. The angle θ between the MultiCam optical axis and the horizontal axis can be calculated as:

$$\theta = \widehat{ABO} + \frac{\widehat{AOB}}{2} = arcsin(\frac{H}{OB}) + \frac{\widehat{AOB}}{2} \tag{5.1}$$

Where H is the height of the MultiCam. $\widehat{AOB}$ is the vertical aperture angle of the MultiCam.

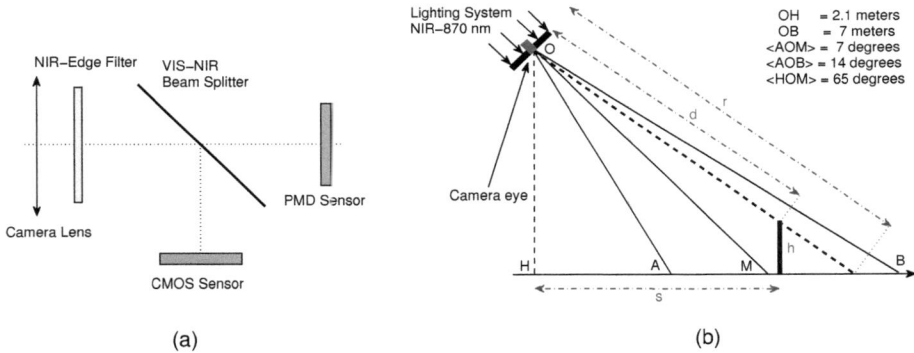

Figure 5.1: (a) Optical set-up of the MultiCam. (b) System set-up.

Hence, the real world distances from all objects in the scene to the MultiCam are lesser than 7.5m. There is no more distance aliasing effect under this system-setup. Also, this setup helps to avoid the direct sunshine to the camera, which consists of a wide range of light spectrum including the visible light, UVA, UVB and even infrared, which affects the distance measurement (see more in [Nguyen et al., 2010a]). Finally, a preferable speed of a reliable autonomous navigation system up to date is at about 3 m/s, so the maximum distance restriction of 7.5m is acceptable.

5. General Vegetation Detection Using an Integrated Vision System

5.2 Spatial Features

As mentioned in Section 5.1 that the 2D-3D setup of the MultiCam is monocular, so an implicit calibration is done by a two dimensional translation function which maps a 10x10 2D pixel to one single PMD pixel. If we assume that the roll and yaw angles are zero, then there is only a pitch angle as shown in the system set-up in **Fig. 5.1(b)**. Whereby, the 3D point cloud can be reconstructed by comparing distances of a query scene with of the flat ground. Indeed, if we assume that $R = \{r_i : i = 1, 2..N\}$ and $D = \{d_i : i = 1, 2..N\}$ are sets of distance values measured from the smooth and level ground surface and from a query scene, respectively. The height of each point in the query scene can be calculated as follows:

$$h_i = H \times \frac{r_i - d_i}{r_i} \qquad (5.2)$$

Similarly the distance of each point in the scene to the vehicle can be computed as:

$$s_i = \sqrt{r_i^2 - H^2} - \sqrt{r_i^2 - H^2} \times \frac{h_i}{H} \qquad (5.3)$$

Each pair of (h_i, s_i) and the position of the point projected into the imagery plane of the 3D sensor provide enough information to reconstruct the 3D point. Then, a 3D scene can be reconstructed by putting the correct colour information into the 3D point cloud. **Fig. 5.2(a)** shows some examples of reconstructed 3D scenes using our approach. Clearly this approach is incompletely performing extrinsic calibration with assumptions of having zero roll and yaw angles as well as perfect smooth and level ground surface to generate the flat ground model, thus, is not very precise. However, a complete and automatic 2D/3D calibration (with intrinsic and extrinsic parameters) returns unstable results due to the low depth resolution (64x48). Indeed, good features are not always obtained to estimate properly intrinsic and extrinsic parameters. The depth image contains noise, thus, affects to the process of estimating intrinsic parameters even if the calibration is done manually. Overall, the proposed method is a clever way to solve the calibration problem in our conditions.

5. GENERAL VEGETATION DETECTION USING AN INTEGRATED VISION SYSTEM

In order to find spatial features, we suggest to over-segment the colour image into many small regions of interest, and map those regions into the depth image to result the corresponding regions of interest. The segmentation technique used in our work is Efficient-Graph-based which was originally introduced in [Felzenszwalb & Huttenlocher, 2004]. To eliminate illumination noise and to save processing time, it is recommended to down-sample the colour image beforehand using Pyramid technique. Thus, the time requested for the segmentation is at around $40ms \rightarrow 60ms$ which is very fast. Then, we want to calculate the spatial features for each small region based on the 3D information reconstructed. An expert in PMD sensor might raise a question that PMD sensor is known to be strongly affected by the sunlight, so is that a good idea to be based on distance information given the camera? We answer "yes" because the strong reflectance of vegetation against near-infrared helps to eliminate illumination noise, thus, provides reasonable depth for the vegetation areas from the scene [Nguyen et al., 2012c]. Consequently, we expect that there should be approximately no wrong distance measurement for vegetation while there usually exists for a non-vegetation area. This insight enables an idea to classify vegetation regions and others based on evaluating their noise and spatial distribution. With the camera positioned as in **Fig. 5.1(b)**, we expect to see small vegetation regions as rough surfaces. So if we build a least square plane for each region, the total sum of point-to-plane distances from all 3D points inside the region can tell us about the smoothness of its surface, denoted by S_{smooth}. On the other hand, in a small region, if the distance of a 3D point to the least square plane is superior to three times of the average of all other points' point-to-plane distances, the 3D point should be a defect. The number of defects found inside a small region can tell us the probability of measurement noise appearing in the region, denoted by P_{noise}. In this work, we will prove that S_{smooth} and P_{noise} are good features to be trained to generate robust vegetation detection classifier. Still we acknowledge that S_{smooth} and P_{noise} could not be used efficiently to detect bushes which are not thick enough due to a "look through" effect where some light beams may pass through the vegetation and return larger ranges.

5. General Vegetation Detection Using an Integrated Vision System

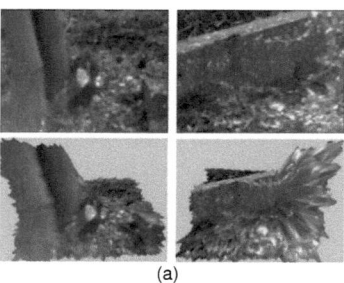
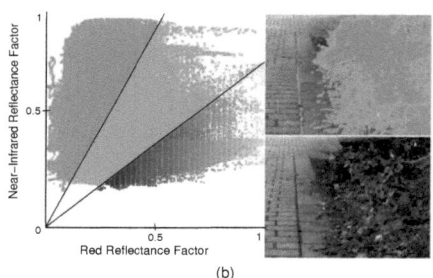

(a) (b)

Figure 5.2: (a) Examples of reconstructed 3D scenes where the exposure-times of 2D and PMD sensors are set at 10 ms. (b) Example of vegetation detection based on thresholding NDVI values where the green colour represents living vegetation, cyan colour denotes dead grass or wet soil. If giving a threshold: $T = \frac{NIR-Red}{NIR+Red} \rightarrow NIR = \frac{1+T}{1-T} Red$, this is a line passing through the origin with the gradient $\frac{1+T}{1-T}$.

5.3 Vegetation Index Calculation

Although, vegetation indices have been widely used in many remote sensing applications to detect vegetation areas, it is still a problematic thought to apply them directly to mobile robotics applications due to the drastically different viewpoints. Regarding to autonomous ground navigation, it would be more challenging in order to deal with views of the sky, shinning, shadow, underexposed, and overexposed effects as well as the presence of a variety of different materials from which the reflected light can have a spectral distribution that is different from that of the sunlight. Intuitively, the sensor noise and errors cause heavy effect in underexposed areas [Bradley et al., 2007], as a result, they are usually confused as vegetation areas. Fortunately, the MultiCam uses an active lighting source to send a modulated NIR signal and receive the reflected NIR signal through the PMD sensor, so that it is not significantly influenced by those illumination effects. The point is that we measure the reflectance of the modulated light from the MultiCam's lighting source instead of sunlight. There might be differences of light spectral distributions, so:

Is it possible to only use NIR reflectance for vegetation detection? or:
Is it still possible to use NDVI for vegetation detection?

5. GENERAL VEGETATION DETECTION USING AN INTEGRATED VISION SYSTEM

The answer is "yes" for both questions in case of good lighting condition and "no" for both when an irregular illumination effect occurs. Since the active light has wavelengths lying within the band from 800nm to 1400 nm (focusing on 870 nm in our case) which is strongly reflected from chlorophyll, the representation of NIR or NDVI maintains its properties regarding to vegetation index. This has been proved practically in our laboratory with 1000 scenes as well as video sequences captured under different lighting conditions and different exposure-times set. In case of only using NIR values, vegetation areas always show very high reflectance from NIR bands compared with others, that means their representation is brighter in the intensity image given by the PMD sensor. Thus, vegetation can be detected by evaluating or thresholding the NIR values. The results of this approach are quite impressive if there is not much light intensity and light colour change of the viewed scenes. In contrast, the threshold must be changed non-linearly upon changes of lighting conditions of surroundings, which degrades the reliability of the approach in outdoor environments.

The use of NDVI is still preferred in our approach, which considers both NIR reflectance and illumination information. However, the NIR and red intensities are given by two different light sources: one comes from the MultiCam, the other comes from the sunlight, which are approximately independent.

Let's recall the normalized difference as follows,

$$NDVI = \frac{NIR - Red}{NIR + Red} \qquad (5.4)$$

From the mathematical point of view, this computation of the difference is more cumbersome than of the simple ratio, including one subtraction, one addition and one division. A faster calculation was proposed in [Crippen, 1990] by a reduction to the ratio between the NIR and the addition of NIR and red, named as infrared percentage vegetation index, which is functionally and linearly equivalent to the NDVI.

$$\frac{NIR}{NIR + Red} = \frac{1}{2} \times \frac{NIR - Red}{NIR + Red} + \frac{1}{2} \qquad (5.5)$$

Consequently, this fast calculation is applied to compute the normalized differ-

5. General Vegetation Detection Using an Integrated Vision System

ence vegetation index in our algorithm. Alternatively, from mathematics point of view, the NDVI is negatively proportional with the red. As a result, these create *the lower luminance level, the higher NDVI*. For instance, very dark areas usually act like chlorophyll-rich vegetation ones due to high values of NDVI (because of small values of red). Hence, giving a threshold or binary linear classification for NDVI values to detect vegetation is not really robust for scenes in complex environments or under illumination effects. An example of detecting vegetation by easily thresholding the NDVI values is described in space of NIR and Red reflectance as separation curves passing through the origin (shown in **Fig. 5.2(b)**). Clearly, there is misdetection between dead grass and soil. Therefore, to avoid biasing NDVI formula, we propose to use brightness descriptor feature in the classifier to compensate changes on red band due to light changes.

5.4 Colour and Texture Descriptors

Previous section has mostly done the work of detecting vegetation. However, there are materials absorbing much NIR light such as water, wet soil, wool, fungus/mould construction and many other artificial ones, which can not be clearly distinguished from vegetation relying on NDVI. Thus, we propose to add colour and texture descriptors to give more confidence for the vegetation detection system. Indeed, vegetation owns its typical colour of green, orange and yellow which are different from that of those NIR absorbing materials.

Colour Analysis: Since the colour of a vegetation area is quite homogeneous, so an easy comparison between a query region's colour and yellow/green/orange colour model can also give reasonable results. For more robustness, we suggest to use histogram distances in order to deal with the intensity and light colour change in outdoor environments. In fact, histogram distances are proved to be discriminative features in image retrieval [Chakravarti & Meng, 2009] [Kumar *et al.*, 2009] [Nguyen *et al.*, 2011b]. In our previous work [Nguyen *et al.*, 2011b], we are able to efficiently detect vegetation in form of regions of interest by computing the histogram

5. GENERAL VEGETATION DETECTION USING AN INTEGRATED VISION SYSTEM

distances between the colour histograms of each region of interest and three colour histogram models. In this work, we also build three colour histogram models (yellow/green/orange) and use the histogram distances as colour features, see more details in. Nevertheless, instead of doing segmentation on the depth image as in [Nguyen et al., 2011b], the colour image is directly segmented to result the image in form of regions of interest (ROIs).

The work of [Jeong et al., 2004] showed the out-performance of Histogram Intersection in HSV colour space compared with it in RGB colour space and with Euclidean Distance in both colour spaces, with respect to image retrieval. However, the work of [Nguyen et al., 2011b] pointed out that the Histogram Quadratic maintains its higher applicability than others' for vegetation detection under different lighting conditions. However, the computation of Histogram Quadratic distance is very expensive, up to 300 ms for such image of 640x480 pixels, which is no longer suited for real-time applications. If the time is not critical, this feature is preferred to give more robust results of vegetation detection.

Texture Analysis: Human eyes can easily recognize vegetation based on texture, which motivates us to investigate texture analysis for vegetation detection. In fact, even with human knowledge of texture analysis, it is still hard to describe the texture of vegetation in general due to a variety of vegetation species which own quite different textures. The interesting insight is that such textures are unstructured in nature compared with artificial materials' because the human tends to design things in a linear structure. To utilize this property, we propose to use Gabor Filter bank, which is well-known to be accurate in estimating texture orientation, to extract texture features. Step by step to obtain Gabor response images from different orientations can be followed by the work of [Kong et al., 2010]. However, to speed up our algorithm we only use 18 orientations instead of 36 as in the work in [Kong et al., 2010]. If [Kong et al., 2010] tried to get long edges to detect lines of road, we in contrast try to remove all edges in the Gabor response images. We obtain 5x18=90 Gabor responses (5 scales and 18 orientations). At each orientation, we compute the average of texture intensity for all scales. Overall, we finally have 18 average Gabor responses (AGRs). Project the ROIs from the over-segmented colour image into AGRs to result small

5. General Vegetation Detection Using an Integrated Vision System

regions of interest there, denote as $G-ROIs$. Observing that: 1) If a small region contains long edges or linear structures, so the maximum texture orientation of the region should be the same as of the edge pixels. 2) If a small region is a smooth surface, so the Gabor responses of the region have very low intensities. 3) If a small region has an unstructured texture, so many edge pixels have different maximum texture orientations as of the region. We define the maximum texture orientation of a small region (or of a pixel) as the orientation (or the angle) at which the Gabor response of the region(or the pixel) is maximized. Hence, for each ROI:

- Calculate the average intensities of corresponding $G-ROIs$. Set *counter* $= 0$.

Figure 5.3: Top-left: colour image; Top-right: segmented image; Bottom-left: unstructured points extracted; Bottom-right: texture map is obtained by weighting the average intensity of Gabor responses by the percentage of unstructured points inside the region.

5. GENERAL VEGETATION DETECTION USING AN INTEGRATED VISION SYSTEM

- If the averages are small, remove the ROI out of interest.

- Else, search for the maximum texture orientation of the region (for example, at the orientation α corresponding to the $G - ROI$ which has the maximum average intensity). For each pixel of the ROI, search for its maximum texture orientation, for example at the orientation θ.

    - if $\alpha = \theta$, the pixel is an edge one.
    - else, the pixel is an unstructured texture one. $counter = counter + 1$

- Percentage of unstructured texture = counter / (total number of pixels inside the region).

Then, ROIs which have less percentage of unstructured texture points are removed. The rest of ROIs are potential to be vegetation regions. Within this consideration, the probability of a ROI to be a vegetation should be positively proportional with the percentage of unstructured texture pixels within the region. Consequently we obtain the texture map as in **Fig. 5.3(b)**. Whereby the unstructured texture regions show brighter intensities than others. We will prove that the intensity of the texture map image can be used as a good texture feature for training vegetation classifier, denoted by $uTexf$.

5.5 Experiments and Results

In order to demonstrate the applicability and reliability of the proposed approach, 1000 scenes and 10 video sequences of outdoor environments were captured by the MultiCam, in both morning and afternoon conditions (see our autonomous ground robot's configuration in [Nguyen et al., 2011b]). Whereby, 500 scenes are used for training and the other 500 for testing. Structured scenes were gathered from sites in the campus of University of Siegen, and cluttered outdoor scenes were taken from the mountain nearby the university. The detection results are evaluated by comparing the output of classifier with hand-labelled data.

5. General Vegetation Detection Using an Integrated Vision System

Table 1 [1] shows the accuracy of vegetation detection results under different features sets trained by binary support vector machine classification algorithm with Radial Basic Kernel [Cortes & Vapnik, 1995] [Lin & Chang, 2011]. The software LIBSVM is available online at [Chang & Lin, 2012]. The times shown in the table are estimated from running the proposed method in the autonomous ground vehicle's computer with Intel Core 2 Dual CPU L7500 2x1.67 GHz and 4 GB of RAM. Clearly, the performance of the approach involves a trade-off between accuracy and speed.

Table 5.1: Confusion Matrices for Different Feature Sets (%)

NDVI,Brightness(Time:128ms)

P-classes/T-classes	Vegetation	Others
Vegetation	76.51	23.49
Others	10.62	89.38

NDVI,Brightness,S_{smooth},P_{noise}(Time:138ms)

P-classes/T-classes	Vegetation	Others
Vegetation	85.08	14.92
Others	8.35	91.65

NDVI,Brightness,S_{smooth},P_{noise} uTexf,HI(Time:416ms)

P-classes/T-classes	Vegetation	Others
Vegetation	94.18	5.82
Others	4.02	95.98

NDVI,Brightness,S_{smooth},P_{noise} uTexf,HQ(Time:1280ms)

P-classes/T-classes	Vegetation	Others
Vegetation	95.10	4.990
Others	2.51	97.49

The more features used the higher detection accuracy achieved. So when time is not critical, it is recommended to use all features. Otherwise, HQ feature is usually ignored due to its expensive computational complexity.

If comparing with the performance of [Nguyen et al., 2011b] (with 85% of precision and time of 2580 ms per frame), our approach is much faster and more robust. Since [Nguyen et al., 2011b] used laser scanner SICK for data acquisition which is extremely time-consuming, the use of NDVI in our approach is more oriented and discriminative for vegetation detection than geometric distribution. Although the proposed approach is less-robust than of [Bradley et al., 2007], it runs much faster and can be used for real-time applications. In fact it is not clear about the processing time of the approach in [Bradley et al., 2007] but we can see that they have to

---

[1] P-classes ≡ Predicted classes; T-classes ≡ True classes; S_{smooth} ≡ Smoothness of surface; P_{noise} ≡ Probability of measurement noise; $uTexf$ ≡ Unstructured texture feature.

5. GENERAL VEGETATION DETECTION USING AN INTEGRATED VISION SYSTEM

use Ladar information which is time-consuming due to the slow data acquisition and the need of a processing time to be interacted with 2D information. Compared with the preliminary version of the paper [Nguyen et al., 2011c], the additional use of unstructured texture and spatial features helps to provide more robust vector components. Consequently, the accuracy improves more than 2% in average with available data. Nevertheless, the processing time is increased additionally of 200 ms per frame. Importantly, we have recognized that without unstructured texture and spatial features, the resulted classifier could not distinguish vegetation from warm objects (human/animal body) or strong NIR reflection objects (green mirror/synthetic clothing/red vehicle paint [Bradley et al., 2007]). Therefore, if we consider many scenes including the presence of those objects, the result of the current approach must be much higher overall accuracy than of the preliminary one. Unfortunately, the evaluation on the performance of the current approach in detecting vegetation with the presence of all those objects has not yet completely done at the current state. We are still in the process of testing the reflectance of different materials in the MultiCam's modulated NIR band under different illuminating conditions. Thus, the future work

Figure 5.4: Examples of vegetation detection results obtained from our approach. The first three images are captured with the camera positioned as in **Fig. 1(b)**, when the robot goes (a) down slope, (b) up slope, (c) on flat road. The last image is captured when the camera is positioned horizontally.

5. General Vegetation Detection Using an Integrated Vision System

should clarify concretely how many percentage the approach achieves in classifying vegetation and warm objects or strong NIR reflection objects.

Finally, **Fig. 5.4** shows some examples of vegetation detection results achieved from our algorithm, which also reflects somehow the range of difficulty of our dataset. Overall, using the feature set described in the **Table 1(Bottom-Left)** seems to be the best choice to balance the accuracy versus computation time, so that the frame-rate can be achieved at about 2.4 fps and the accuracy at 95.08% ($= \frac{95.98+94.18}{2}$, the number of positive and negative samples are equal).

5.6 Conclusion

We have introduced an efficient approach for vegetation detection using a Multi-Cam which is mounted both 2D and PMD sensors into a monocular set-up. The benefit is to have a sufficient 2D-3D information from a single vision device. The achievements of the technical research in this paper are to provide the optimal features used to robust the vegetation detection classifier. Whereby, the results have been demonstrated to be robust as well as the consuming time is short, which proves that the proposed approach can be used for on-board navigation. Image databases and some videos of demonstrations are available online at [Nguyen, 2012]. Remarkably, the paper enables a possibility of distinguishing vegetation from warm objects (human/animal body) or strong NIR reflection objects (green mirror/synthetic clothing/red vehicle paint) by using texture and spatial feature descriptors in the classifier, which is infeasible in previous approaches. Overall, our method outperforms conventional approaches concerning high precision and fast processing. Still the approach has not been fully tested to be very robust in the presence of warm objects or strong reflection objects, so the future work should carry out a concrete evaluation on the performance of the approach on classifying vegetation and those objects.

5. GENERAL VEGETATION DETECTION USING AN INTEGRATED VISION SYSTEM

Chapter 6

Spreading Algorithm for Efficient Vegetation Detection

The use of the MultiCam in detecting vegetation has been investigated successfully in Chapter 3 and Chapter 5. The use of an independent lighting source helps to stabilise NIR reflectance, thus, reducing the hue impact of light changes.

In Chapter 3, a new vegetation index, the so called modification of normalized difference vegetation index, has been derived to impressively detect vegetation. Generally, the approach is fast and robust. However, it fails to deal with the presence of warm or strong reflection objects. Also, the MNDVI is not applicable in dim lighting conditions.

In Chapter 5, a classification-based method has been introduced to detect general vegetation. The accuracy is high when a set of vegetation indices, spatial and visual features is used to train vegetation classifier. Remarkably, the addition of colour and texture features into the feature vector potentially helps to distinguish vegetation from warm or strong reflection objects. However, many features need to be extracted, and then trained, which degrades the usability of the method for real-time applications, especially for on-board navigation with high speed.

Therefore, this chapter studies a way to improve the robustness of the multi-spectral approach in Chapter 3 and the speed of the classification-based method in Chapter 5. This leads to an idea of detecting chlorophyll-rich vegetation using the

6. SPREADING ALGORITHM FOR EFFICIENT VEGETATION DETECTION

multi-spectral approach, and a spreading algorithm would help to spread out the vegetation based on colour and texture. Thus, the aim of this chapter is to create an adaptive learning algorithm which performs a quantitatively accurate detection that is fast enough for a real-time application. Indeed, chlorophyll-rich vegetation pixels are selected by thresholding vegetation indices, and then considered as the seeds of a "spread vegetation". For each seed pixel, a convex combination of colour and texture dissimilarities is used to infer the difference between the pixel and its neighbours. The convex combination, trained via semi-supervised learning, models either the difference of vegetation pixels or the difference between a vegetation pixel and a non-vegetation pixel, and thus allows a greedy decision-making process to expand the spread vegetation, the so-called vision-based spreading. To avoid overspreading, especially in the case of noise, a spreading scale is set. On the other hand, another vegetation spreading based on spectral reflectance is carried out in parallel. Finally, the intersection part resulting from both the vision-based and spectral reflectance-based vegetation spreading is added to the spread vegetation. The approach takes into account both vision and chlorophyll light absorption properties. This enables the algorithm to capture much more detailed vegetation features than does prior art, and also give a much richer experience in the interpretation of vegetation representation, even for scenes with significant overexposure or underexposure as well as with the presence of shadow and sunshine. In all real-world experiments we carried out, our approach yields a detection accuracy of over 90%, which outperforms conventional approaches.

This work has been published in **Journal of Robotics and Autonomous System** [Nguyen *et al.*, 2012b].

6.1 Introduction

The repeated occurrence of task rejection of autonomous robots during forest exploration in the European Land Robot Trial (ELROB) from 2005 to 2011 indicated that a basic task such as obstacle avoidance could become an ever challenging issue in a

6. Spreading Algorithm for Efficient Vegetation Detection

cluttered outdoor environment, especially with the presence of vegetation. Indeed, the concept of a lethal obstacle simply defined as a solid and significantly high object is no longer applicable for vegetation such as tall grass and leaves. Otherwise, unnecessary obstacle avoidance operations of the robot would drive it to a situation of losing its way or stopping due to all paths being blocked by dense geometric obstacles. Therefore, safe and reliable autonomous navigation requires a growing need of an efficient vegetation detection module integrated in every autonomous mobile outdoor robot. Locating vegetation areas in a scene helps not only to determine which traversable way to pass but also to understand the local environment for a re-allocation purpose worthy of use in the case of Global Positioning System (GPS) loss. Furthermore, driving on grass or leaves for example would increase wheel slippage, which causes errors in the odometry. Hence, vegetation detection lets the robot know which type of terrains it is dealing with, and thus which strategies should be applied.

Upon seeing the image in **Fig. 6.1**, a human has no difficulty in understanding its use of colour and texture to point out vegetation areas. However, inferring specific properties of vegetation in general remains extremely challenging for current computer vision systems. Different species of vegetation have different shapes, textures, structures and colours. Thus, previous works in vegetation detection focused on several specific types of vegetation such as specific colour: green leaf, green grass [Gu & Zhong, 210]; specific structure: foliage, needle tree [Lalonde et al., 2006]; specific texture: grass-field [Zafarifar & de With, 2008], weed [Sabeenian & Palanisamy, 2009]. Even limiting applicable species, algorithms relying on only

Figure 6.1: From left to right: an original image; near-infrared image; texture image created by the prosed approach; vegetation marked by the proposed algorithm.

6. SPREADING ALGORITHM FOR EFFICIENT VEGETATION DETECTION

vision features often end up ignoring illumination effects including shadow, strong shining, underexposure and overexposure which are inevitable outside. As a result, those approaches are not stable and reliable enough for use in a safe navigation system.

Recently, [Lalonde et al., 2006] [Lu et al., 2009] and [Nguyen et al., 2010b] presented algorithms for analysing 3D structures of foliage-like vegetation from 3D point clouds captured by a laser scanner. [Wurm et al., 2009] measured the remission of a laser to classify vegetation and non-vegetation regions in a structured environment. [Nguyen et al., 2011b] and [Lu et al., 2009] proposed 2D-3D feature fusion approaches combining colour, texture and 3D distribution information to detect vegetation. Although the use of laser data can improve the stability against illumination changes, it significantly slowdowns the detection rate due to the long time needed for data acquisition by a common laser scanner (exceptionally Velodyne LIDAR is fast and robust but quite expensive).

Regarding the photosynthesis of vegetation, visible light, especially red or blue light, is strongly absorbed by the chlorophyll in vegetation. The cell structure of the leaves, on the other hand, strongly reflects near-infrared light (from 0.7 to 1.1 μm). Therefore, vegetation indices established by measuring the ratio of radiances in the near-infrared (NIR) and red bands can be used to detect vegetation, for instance detecting the green surface of the earth in the remote sensing field [Tarpley et al., 1984] [Townshend et al., 1985] [Crippen, 1990] [Tucker et al., 1986]. Surprisingly, there is not much investigation available on utilizing this promising property for ground-based terrain classification for navigation. Remarkably, one of the few contributions successfully exploiting the spectral reflectance of vegetation was introduced by [Bradley et al., 2007]. The author, however, still needs to additionally use laser data to approach a more robust vegetation detection. Interestingly, [Nguyen et al., 2012c] has shown that by varying the exposure time and adding independent light, the vegetation detection system performs in a more robust and stable way against illumination changes. Nevertheless, such a system could not work well in dim light condition (see the explanation in section 6.2). Also, based only thresholding vegetation indices, the approaches in [Bradley et al., 2007] and [Nguyen et al.,

6. Spreading Algorithm for Efficient Vegetation Detection

2012c] could not lead to a complete solution for an automatic vegetation detection in really different light conditions. A manual adjustment on the range of the vegetation index threshold is usually required. To overcome these difficulties, this paper introduces a spreading algorithm to automatically detect vegetation in cluttered outdoor environments. This paper is a follow-up on the system setup used in [Nguyen et al., 2012c].

The goal of the paper is to create an adaptive learning algorithm which performs a quantitatively accurate detection that is fast enough for a real-time application. We use the insight that every vegetation pixel should have a significant reflectance from NIR as well as strong absorption of visible light, and that two adjacent vegetation pixels should have very similar colours and textures. So, we simply detect chlorophyll-rich vegetation by setting high thresholds on vegetation indices including Normalized Vegetation Index [Bradley et al., 2007] and Modification of Normalized Vegetation Index [Nguyen et al., 2012c]. The chlorophyll-rich vegetation pixels are then considered as seeds of our "spread vegetation". We spread out the spread vegetation based on visual difference and spectral reflectance difference in parallel. The intersection part between vision-based and spectral reflectance-based vegetation spreading is then judged as vegetation.

Remarkably, instead of building colour models for vegetation, the colour similarity measure between vegetation seeds and the neighbouring pixels, in order to expand the "spread vegetation", helps to deal with a variety of vegetation appearing in different colours. The novelty of the paper also lies in the finding of unstructured texture points extracted from analysing the texture orientation of a colour image, which helps to distinguish vegetation from other artificial objects with dense edges.

The paper is organized as follows. First, the spectral reflectance of vegetation is investigated in a more detailed manner than in previous work in order to deal with illumination changes, explained in section 6.2. Second, we introduce new visible features including colour and texture which will be then proved to be suited for representing characteristics of general vegetation, see section 6.3. Section 6.4 describes the spreading algorithm for detecting and grouping vegetation. Experiments and results are illustrated in section 6.5 while section 6.6 concludes this work.

6. SPREADING ALGORITHM FOR EFFICIENT VEGETATION DETECTION

6.2 Discussion on Vegetation Indices

The NDVI has been proved to be problematic in oversaturation and underexposure conditions in [Nguyen et al., 2012c]; this is because the light absorption spectrum of all objects including vegetation changes considerably. In order to be more stable against light changes, we proposed using a MultiCam equipped with a NIR lighting system (wavelengths centred at 870 nm) in Nguyen et al. [2012c]. The MultiCam integrates Photo Mixer Device (PMD) and CMOS sensors into a molecular setup, and thus provides simultaneously NIR and colour images with resolutions of 64x48 pixels and 640x480 pixels, respectively. The intensity of the lighting source as well as the gain of PMD sensor are adjustable, which helps to stabilize the received NIR intensity. Then, the gain of the CMOS sensor is slaved to match the gain of the PMD sensor plus a constant offset. In that way, we obtain a more stable multi-spectral system. In order to test the changes of the NIR and red bands against illumination effects, a measure of changes in the near-infrared and red bands in terms of luminance has been done in part of this work. Concretely, we took 500 scenes of vegetation in different illuminating conditions and normalized the NIR, red and luminance information before doing regression analysis for the relationship between NIR/red and luminance. As a result, the changes are most likely in a logarithm form instead of a linear one, wherein the change of the red band is much stronger than of the NIR band, especially in the case of overexposure or strong shining conditions, while it is lower in the case of underexposure or shadow, see more in [Nguyen et al., 2012c]. [Nguyen et al., 2012c] also proved that Normalized Difference Vegetation Index should be modified as follows when using an active NIR lighting system:

$$MNDVI = \frac{NIR - log(Red + \varepsilon)}{NIR + log(Red + \varepsilon)} \quad (6.1)$$

where $\varepsilon = 1$ is used to guarantee a positive value of the log, so this index ranges from 0 to 1.

Generally, the MNDVI shows better performance compared with the NDVI in detecting vegetation under different lighting conditions (see **Fig. 6.2a,b,c**) (see also

6. Spreading Algorithm for Efficient Vegetation Detection

the comparison between the two indices in [Nguyen *et al.*, 2012c]). However, the softening impact of Red in the MNDVI causes missed detection of some species of vegetation which absorb the red band from the sunlight very strongly but reflects less near-infrared light, especially in circumstances of underexposure or dim light conditions; see **Fig. 6.2d,e**.

Proof: Simply, when the value of Red is quite small, we have $log(Red+1) \approx 0$, so $MNDVI \approx 1 \; \forall \; NIR$.

In this paper, we propose a possible combination between the two indices to result in a more robust detection using a MultiCam. Intuitively, the NDVI and MNDVI can supplement each other to create a more stable vegetation detection against illumination effects; see **Fig. 6.2**. Indeed, we will prove that the supplementation is worthy of the aim of detecting vegetation in a cluttered outdoor environment under illumination changes using our algorithm.

6.3 Visual Features for Scene Understanding

There are many thousands of vegetation species available around us, which have different shapes, colours, and textures in different lighting conditions. Therefore, the work of finding a common characteristics of vegetation based on vision is difficult. However, the question why a human can without doubt easily recognise vegetation motivates us to come back again to learn visible features in order to generate possible discriminative features of vegetation. Actually, the human eye sees different plant leaves as shades of green/red/orange/yellow etc., as characterized by the corresponding colour peaks in reflectance spectra. The eye/brain colour system can differentiate shades of the colours under different lighting conditions, which is still infeasible or too complex for current computer vision systems. Instead of giving a specific colour model, this work prefers using a convex combination of colour and texture dissimilarities to infer the visual difference, which partially relies on the property 1.

Property 1: Although different species of vegetation can have different colours, the colour in a small vegetation region is expected to be homogeneous.

6.3. VISUAL FEATURES FOR SCENE UNDERSTANDING

The property enables the idea that, if we know a vegetation pixel, we just need to search for a connected vegetation pixel among neighbours which have very similar colours. Thus, we choose colour dissimilarity as one distance parameter for estimating the visual difference in our algorithm. In order to be invariant against illumination changes, we use the colour dissimilarity in the opponent colour space which is explained in the next subsection.

6.3.1 Opponent Color Space

Colour dissimilarity has no invariance properties in the RGB colour space. Similar to the work of [van de Sande et al., 2010], we rotate the RGB colour space and then swap two channels R and G so that intensity changes do not affect the colour information in the new colour space, the so called opponent colour space (see Fig. 6.3).

In terms of mathematical expression, the transformation can be written as follows:

$$\begin{pmatrix} O_1 \\ O_2 \\ O_3 \end{pmatrix} = \begin{pmatrix} \frac{G-R}{\sqrt{2}} \\ \frac{G+R-2B}{\sqrt{6}} \\ \frac{G+R+B}{\sqrt{3}} \end{pmatrix} \quad (6.2)$$

The reason to swap the two channels is to have a positive relation between the O_1 and Green channels. We also can see that O_1 and O_2 represent colour information while O_3 denotes intensity information. We will prove that the two pieces of colour information (O_1, O_2) are invariant to an intensity shift.

Proof: Assume an image I has its RGB colour space and its opponent colour space (O_1, O_2, O_3). I' is a shifted version of I, where $R' = R + \delta$; $G' = G + \delta$; $B' = B + \delta$. The opponent colour space of I' can be written as

$$\begin{pmatrix} O'_1 \\ O'_2 \\ O'_3 \end{pmatrix} = \begin{pmatrix} \frac{G'-R'}{\sqrt{2}} \\ \frac{G'+R'-2B'}{\sqrt{6}} \\ \frac{G'+R'+B'}{\sqrt{3}} \end{pmatrix} = \begin{pmatrix} \frac{G+\delta-R-\delta}{\sqrt{2}} \\ \frac{G+\delta+R+\delta-2B-2\delta}{\sqrt{6}} \\ \frac{G+\delta R+\delta+B+\delta}{\sqrt{3}} \end{pmatrix} \quad (6.3)$$

6.3. Visual Features for Scene Understanding

Hence, $O'_1 = O_1$ and $O'_2 = O_2 \rightarrow$ intensity shift invariant. Certainly, when the intensity shifts between different channels (red/green/blue) are different from each other (or different δ), the proposed method also could not result in any invariance property.

The colour difference between two pixels $(O_1[i], O_2[i])$ and $(O_1[j], O_2[j])$ is computed as follows:

$$C_{i,j} = \sqrt{(O_1[i] - O_1[j])^2 + (O_2[i] - O_2[j])^2} \quad (6.4)$$

Other than trying to detect a specific colour of vegetation, our approach enables more chances to detect varieties of vegetation which can have many different colours. In fact, this colour dissimilarity measure is not invariant against illumination changes. However, the algorithm still works well in many cases of illumination changes because such a colour dissimilarity measure does not change very much or two adjacent vegetation pixels are still expected to have similar colours. We acknowledge that more investigation on colour invariance as in the works Finlayson et al. [2006] and Berens & Finlayson [2000] can improve the performance of the algorithm. This should be taken into account in our future work.

6.3.2 Unstructured Texture

Intuitively, vegetation has a texture. However, to define the texture in general or find a characteristic of the texture which can differentiate one type of vegetation from another is often impossible even for a human. So, our goal is not to find a discriminative texture of vegetation but a common characteristic that most of types of vegetation share. Commonly, current approaches use edge detection methods to estimate the complexity of image textures as well as infer the textures. Nevertheless, how to distinguish between vegetation and another object with dense edges is still really difficult. Investigating more on that issue we recognise that, from a human eye's perception, most types of vegetations have unstructured textures.

Property 2: The textures of most types of vegetation are unstructured or turbulent.

6.3. VISUAL FEATURES FOR SCENE UNDERSTANDING

This can be inferred as we would find many pixels in a small vegetation region, which have different texture orientations from the texture orientation of the region.

This is a very interesting finding, because we can eliminate edges which emphasize the texture orientation of a small local region. This distinguishes between a texture of vegetation/unstructured soil and of artificial objects with dense edges. The following explains how to exploit this property to find a good texture feature of vegetation.

Since Gabor filters are known to be accurate in estimating texture orientations [Rasmussen, 2004], which has been shown to be well applied in detecting long edges and structured orientations in the work of Kong et al. [Kong et al., 2010]. In contrast, we are going to prove that Gabor filters can also be used to detect unstructured textures or remove long edges and structured orientations in a similar way. For an orientation ϕ and a scale ω, let $g(\omega,\phi,x,y)$ be the function defining a Gabor filter centred at the origin.

$$g(\omega,\phi,x,y) = e^{-\frac{x^2+y^2}{\lambda^2}} e^{2i\pi\omega(x\cos\phi+y\sin\phi+\psi)} \tag{6.5}$$

Where $\psi = 90$, kernel size = 11, $\lambda = 4.73$. We consider 5 scales($\omega = \omega_0 \times 2 \times k$, $k = 1,2,3,4,5$) on a geometric grid and 6 orientations(180 divided by 30) (see **Fig. 6.4**).

Assume $I(x,y)$ as a gray level value of an image at (x,y). The response of a Gabor filter at a scale ω and orientation ϕ with input I is defined as follows:

$$G = I \oplus g_{\omega,\phi} \tag{6.6}$$

The convolution $\oplus$ returns a real part and an imaginary part, which are then subjected to a square norm to produce the texture intensity.

$$I_{\omega,\phi} = Re(G_{\omega,\phi})^2 + Im(G_{\omega,\phi})^2 \tag{6.7}$$

For saving computational effort, each response image for an orientation is defined

6.3. Visual Features for Scene Understanding

as the average of the responses at different scales.

$$\overline{I_\phi} = \frac{1}{M} \sum_{\omega=1}^{M} I_{\omega,\phi} \tag{6.8}$$

where M = 5 (five scales used in our case). Then, at each position (x,y), we should obtain an expectation vector of the Gabor responses as

$$E_{x,y}^T = [E_0 E_1 .. E_{N-1}] \tag{6.9}$$

The vector includes N elements corresponding to N orientations (= 6 orientations in our case). We assume that θ is the maximum texture orientation of the pixel at position (x,y); then $E_\theta \geq E_\phi \; \forall \phi : 0 \leq \theta, \phi < N$. We rewrite the expectation vector $E_{x,y}$ into a new order if $\theta \geq 1$ (actually we are doing a circular shift),

$$E_{pix}^T = [E_\theta E_{\theta+1} .. E_{N-1} E_0 .. E_{\theta-1}] \tag{6.10}$$

E_{pix} is the so called expectation vector of the Gabor responses at position (x,y), which reflects the texture orientation of the pixel.

In a small region R_s which contains the pixel at (x,y), we compute the average of each expectation element.

$$\overline{E}_\phi = \frac{\sum_{(x,y) \in R_s} E_{x,y,\phi}}{\sum_{(x,y) \in R_s} 1} \tag{6.11}$$

Assume that ξ is the maximum texture orientation of the region, so $\overline{E}_\xi \geq \overline{E}_\phi \; \forall \phi$, $0 \leq \xi, \phi < N$. We rewrite the expectation vector $E_{x,y}$ into a new order if $\xi \geq 1$:

$$E_{reg}^T = [E_\xi E_{\xi+1} .. E_{N-1} E_0 .. E_{\xi-1}] \tag{6.12}$$

E_{reg} is the so called expectation vector of the Gabor responses at position (x,y), which reflects the texture orientation of the region R_s. Therefore, the distance be-

6.3. VISUAL FEATURES FOR SCENE UNDERSTANDING

tween the two vectors represents the turbulent texture property of the pixel at (x,y).

$$T_{d(x,y)} = ||E_{pix} - E_{reg}||_{x,y} \tag{6.13}$$

or

$$T_{d(x,y)} = \sqrt{\frac{(E_\theta - E_\xi)^2 + (E_{\theta+1} - E_{\xi+1})^2 + .. + (E_{\theta-1} - E_{\xi-1})^2}{N}} \tag{6.14}$$

Clearly, if the pixel at (x,y) belongs to an edge, its texture orientation should be the same as the texture orientation of R_s or $\theta = \xi$, so $E_{pix} = E_{reg}$ or $T_{d(x,y)} = 0$. This explains how we distinguish edge points from vegetation points.

Property 3: The distance $T_{d(x,y)}$ is rotation invariant.

Proof: Assume $I(x,y)'$ is a rotated version of $I(x,y)$ and assume the expectation of texture intensity as $E'_{x,y} = \{E'_0...E'_{\theta'}...E'_{N-1}\}$, θ': the maximum orientation of E'. We rearrange $E'_{pix} = [E'_{\theta'} E'_{\theta'+1}...E'_{N-1} E'_0...E'_{\theta'-1}]$. We have rotation properties as follows: $E'_{\theta'} = E_\theta$, $E'_\phi = E_{\phi+\theta-\theta'}$, $E_\phi = E_{N+\phi}$ $\forall \phi$. Hence: $E'_{\theta'+i} = E_{\theta'+\theta-\theta'+i} = E_{\theta+i}$ and $E'_\phi = E_{\phi+\theta-\theta'+N}$ subject to $i = 0 \to N - \theta' - 1$, $\forall \phi$. Or $E'_{pix} = E_{pix}$, rotation invariance.

The confidence to be an unstructured region of the region R_s is estimated by counting the number of turbulent texture pixels over the total number of pixels inside the region.

$$T_{R_s} = \frac{\sum_{T_{d(x,y)} \geq T_0, \forall (x,y) \in R_s} 1}{\sum_{(x,y) \in R_s} 1} \tag{6.15}$$

Thus, if a pixel i belongs to the small region R_s and a pixel j belongs to the small region R_f, then the texture dissimilarity between i and j is estimated as

$$T_{i,j} = T_{R_s} - T_{R_f} \tag{6.16}$$

6. Spreading Algorithm for Efficient Vegetation Detection

A small local region R_s can be defined in many different ways depending on different purposes, for instance, as a small region inside a small rectangular window. In our case, we prefer to semantically connect the texture and colour features, so we define each R_s as a superpixel of the corresponding segmented colour image, similarly to the method of [Saxena et al., 2009] and [Felzenszwalb & Huttenlocher, 2004].

Examples of unstructured texture image response resulting from the above method are illustrated in **Fig. 6.5**. The fourth column of the **Fig. 6.5** shows the confidence maps regarding unstructured texture estimation where brighter means more confidence; see also the confidence expression (**Eq. 6.15**). We can see that most of edges and structured textures have been removed in those examples; the regions of vegetation are highly visible by their unstructured textures.

6.4 Spreading Algorithm

Our algorithm begins by thresholding the Normalized Vegetation Index (NDVI) and the Modification of Normalized Vegetation Index(MNDVI) (see more in section 2) to select chlorophyll-rich vegetation pixels which are then considered as the seeds of our spread vegetation. For each seed pixel, we calculate the distances of colour and texture between the pixel and its neighbours. The neighbours whose distances are smaller than the internal difference of the seed pixel are considered as a set of vision-based spread vegetation. On the other hand, decreasing the NDVI and MNDVI thresholds provides a set of spectral reflectance-based spread vegetation. Pixels belonging to the intersection of the two sets are joined into the spread vegetation . The process is repeated until the percentage of joined pixels over the total pixels of the two sets is less than an epsilon or a tolerance parameter.

A pixel at (x,y) in the image has four dimensions: colour, texture, NDVI and MNDVI. In order to start the spreading algorithm, we need to initialize some hand-tuned parameters: τ_{up}, η_{up}, respectively, are the NDVI and MNDVI thresholds to select chlorophyll-rich vegetation; τ_{low}, η_{low}, respectively, are the lower bounds of

6. SPREADING ALGORITHM FOR EFFICIENT VEGETATION DETECTION

the NDVI and MNDVI thresholds to classify chlorophyll-less vegetation and non-vegetation. To estimate the first two thresholds, we hand-labeled 2000 superpixels as vegetation samples from segmented colour images and manually classified them into chlorophyll-rich vegetation (CRV) and chlorophyll-less vegetation(CLV). A linear classification of the two groups CRV and CLV helps to determine the values of τ_{up}, η_{up}. The values of τ_{low} and η_{low} are obtained from the work of [Bradley et al., 2007] and [Nguyen et al., 2012c] and practical experiments (in our case $\tau_0 = 0.429$, $\eta_0 = 0.961$, $\tau_{low} = 0.231$, $\eta_{low} = 0.956$).

Definition 1: A pixel p is called the parent of a pixel i if and only if p belongs to the spread vegetation and i is connected or joined to the spread vegetation through p.

Algorithm ⎯⎯⎯⎯⎯⎯⎯⎯⎯⎯⎯⎯

1. Thresholding NDVI and MNDVI to select chlorophyll-rich vegetation considered as the seeds of the spread vegetation (sVeg). $\tau = \tau_{up}$ and $\eta = \eta_{up}$

$$SEEDS = \{(x,y) : NDVI_{x,y} \geq \tau \ \&\& \ MNDVI_{x,y} \geq \eta\}$$
$$sVeg = SEEDS$$

2. Segment the colour image into superpixels which are then used for calculating texture features.

3. Assume that ζ is the size of **sVeg**. Repeat step 4 for i = 1,2..ζ.

4. Find the pixel spread by the vision-based approach (see **Fig. 6.6**). At the *ith* pixel of the tree, a pixel j is a neighbour of pixel i in the colour image. Compute the colour distance $C_{i,j}$ using **Eq. 6.4** and the unstructured texture distance $T_{i,j}$ using **Eq. 6.16**. The dissimilarity between the two pixels is defined as

$$D_{i,j} = w_1 \times T_{i,j} + w_2 \times C_{ij} \qquad (6.17)$$

where $w_1 + w_2 = 1$. $D_{i,j}$ is actually a convex combination of colour and texture dissimilarities. Assume that p is the parent pixel of i (see **Definition 1**). If i has no parent then $D_{i,p} = 1/\zeta$.

6. Spreading Algorithm for Efficient Vegetation Detection

- If j does not belong to **sVeg**, the pixel j is added to **sVeg** via the vision-based spreading (VbS) if and only if the dissimilarity between j and i is equal to or smaller than the dissimilarity between i and p, or $D_{i,j} \leq D_{i,p}$.
- If j belongs to **sVeg**, we certainly have $D_{i,j} \leq D_{i,p}$; this constraint is used to optimize w_1 and w_2.

If added:
$$\begin{cases} D_{ij} = D_{ij} + c/\zeta \\ \zeta = \zeta + 1 \end{cases}$$

The constant c represents the spreading scale. The higher the value of c, the more aggressive is the greedy algorithm which lets the "spread vegetation" spread out more aggressively.

5. Find the pixel spread by the spectral reflectance-based approach.

$$\tau = (\tau + \tau_{low})/2$$
$$\eta = (\eta + \eta_{low})/2$$
$$SRbS = \{(x,y) : NDVI_{x,y} \geq \tau \;\&\&\; MNDVI_{x,y} \geq \eta\}$$

6. The intersection part of the both vision-based and spectral reflectance-based spreading is added into the tree: **sVeg** = **sVeg** + ($VbS \cap SRbS$)

7. Estimate the percentage of the pixels added over the total number of candidates given by both vision-based and spectral reflectance-based spreading. Repeat from the step 3 until the percentage is equal to or lower than a tolerance parameter.

Step-by-Step Explanation

As stated in the works of Bradley et al. [2007] and Nguyen et al. [2012c], vegetation pixels lie in the top-left part in a NIR-red space while those of soil and man-made structures are in the middle and those of the sky are at the bottom right. Consequently, it is hard to differentiate and classify between chlorophyll-less vegetation, soil and man-made structures but classifying chlorophyll-rich vegetation is simply

6. SPREADING ALGORITHM FOR EFFICIENT VEGETATION DETECTION

done by thresholding the NDVI and MNDVI. This confirms the validity of step 1 in our algorithm.

Step 2 oversegments the colour image to result homogeneous regions or superpixels. The belief is that pixels inside an oversegmented region of an artificial object should have similar colours and the same texture orientation while pixels inside an oversegmented region of vegetation could have also similar colours but many different texture orientations. This supports the idea of measuring the unstructured texture confidence in step 4. Either vegetation or soil appears as an unstructured texture object, so in order to distinguish vegetation and soil, we combined both texture and colour features. Therefore, step 4 measures the difference between two neighbour pixels based on weighted colour and texture dissimilarities.

Property 4: The dissimilarity measure in step 2 is intensity-shift invariant.

*Proof: The colour dissimilarity is measured in the opponent colour space in order to obtain the intensity-shift invariant property. The measure of unstructured texture confidence is based on the turbulent orientation property of vegetation pixels, which is invariant to the texture intensity shift, and thus also invariant to the image intensity shift (see **Eq. 6.16**). Thus a linear combination between the colour and unstructured texture confidence dissimilarities (see **Eq. 6.17**) results in an intensity-shift invariant property for the dissimilarity measure.*

In step 4, the two pixels (i, j), (i, p) are neighbours. In the case that j is a vegetation pixel, we expect that the dissimilarities of (i, p) and of (i, j) should be very similar (note: i, p are vegetation pixels). Therefore the dissimilarity of (i, j) should be equal or smaller than the dissimilarity of (i, p) plus a spreading scale c/ζ. A higher value of c allows more greedy spreading of vegetation. Practical assessment reveals that the value of c should be 2.4 for scenes with good lighting conditions. The value of c should be decreased (to 1.8) in the case of strong shining or shadow conditions in order to avoid an over-spreading. Thus, an early step of checking lighting conditions of the current scene can guide the choice of the value of c, but this is not so critical.

In step 5, one can easily see that in the case of good lighting conditions, the decreasing τ and η step by step is equivalent to directly setting the lower bounds

$\tau = \tau_{low}$ and $\eta = \eta_{low}$. The aim of that step is just to counter the influence of overspreading via the vision-based spread under extremely strong shining conditions or shadows or extreme overexposure or underexposure. Indeed, when such conditions occur, the performance of the classification based on colour and texture degrades sharply, and thus leads to an overspreading via the vision-based spread. In those cases, it is better to only detect chlorophyll-rich vegetation because the chlorophyll-less vegetation is easily confused with other objects or materials, especially soil. While the decrease of the two thresholds still keeps detecting only chlorophyll-rich vegetation, the vision-based spreading is really greedy. Consequently the "break" in step 7 would help to stop the algorithm. The detection results in this case only rely on the spectral reflectance-based spread.

Examples of the two phases of spreading vegetation are illustrated in **Fig. 6.7**. Clearly, we can see that a single phase has a lot of either false negatives or false positives. The intersection of the two masks tends to be robustly covering vegetation areas.

6.5 Experiments and Results

In this work, we used a MultiCam mounted on the front of an autonomous mobile outdoor robot(see **Fig. 6.8**) to capture near-infrared and colour images. The MultiCam integrates a Time-of-Flight (ToF) sensor and CMOS sensor into a molecular setup. A beam splitter is used to separate the visible light and near-infrared light which are then sensed by the CMOS and ToF sensors to result in colour and near-infrared images, respectively. The configuration of the MultiCam is described in detail in [Ghobadi *et al.*, 2010] [Ghobadi *et al.*, 2008] [Nguyen *et al.*, 2012c]. Actually, this work does not particularly require the use of a MultiCam but a multi-spectral camera with an active NIR lighting source to obtain NIR and colour images as well as stabilise the NIR reflectance. The setup of two mono-cameras with one covered by a NIR-Transmitting filter like in the work of Bradley Bradley *et al.* [2007] is also suitable, but this needs an additional NIR lighting system. Indeed, in the case of not

6. SPREADING ALGORITHM FOR EFFICIENT VEGETATION DETECTION

using an active light source, the MNDVI is no longer useful because the log function in the MNDVI actually makes it more sensitive to changes in illumination intensity rather than less so. We implemented the approach of [Lalonde et al., 2006] using the SICK laser scanner LMS221. The approach of Nguyen et al. [2011b] was carried out using the SICK laser scanner LMS221 and Siemens C810. The approach of [Bradley et al., 2007] was realized by using stereo cameras with one camera covered by a NIR-Transmitting filter and the other covered by a NIR-Blocking filter (we use those filters from Hoya company:http://www.hoyaoptics.com).

Intuitively, the performance of the proposed algorithm is illustrated through **Fig. 6.9** where we consider all illumination effects including shadow, shining, overexposure, underexposure. Looking at the texture feature images in the third row, we can again see that the long edges have been mostly removed, the remaining texture areas mainly consist of turbulent texture points. The vegetation detection is quite robust under various illumination effects; see the last row. Regarding current multi-spectral approaches, a warm object or human is usually confused as vegetation due to high infrared radiation emission. Remarkably, the algorithm can help to classify humans and vegetation in most cases, that is infeasible in previous approaches [Bradley et al., 2007][Nguyen et al., 2012c]. However, in the case that a human is wearing clothes in a very similar colour as vegetation and is staying inside the vegetation area, the algorithm also fails in the classification.

In order to give a quantitative comparison between the proposed method and previous ones, 2000 raw images and 10 videos of outdoor scenes were taken under both morning and afternoon conditions as well as with shadow, shining, underexposure and overexposure effects being taken into account. All data was collected and stored in the robot's computer when the robot traversed throughout outdoor environments. Scenes were then manually classified into five groups: good light condition, underexposure, overexposure, strong shining and shadow. We would like to evaluate the performance of different algorithms for different groups of scenes. For that aim, in each group, we oversegmented colour images into homogeneous regions using superpixel-based segmentation [Felzenszwalb & Huttenlocher, 2004]. Those regions are then manually classified to positive and negative samples. A positive sample is

6. Spreading Algorithm for Efficient Vegetation Detection

understood as a vegetation region while a negative sample is understood as a non-vegetation region. A precision map was established to store the information of each sample including the order of the image to which the sample belongs, the order of the sample when segmented, and the status of the sample: 1 for a positive and 0 for a negative. Therefore, whenever an image is inputted into the algorithm, the outcome will be evaluated by counting the number of correct/incorrect positive samples and of correct/incorrect negative samples compared with the precision map (or ground-truth). **Table 6.1** shows the confusion matrices of different methods for different groups of scenes inputted. In fact, the algorithm based on Local Point Statistic Analysis [Lalonde et al., 2006] does not depend on the illumination changes due to stable laser data. The different detection rates just imply the different complexities of the scenes in an outdoor environment where the group "Shining" contains a few more complex scenes (51.02% true positive percentage) and the group "OverExposure" owns a few less complex scenes (45.52% true positive percentage), with regard to 3D structures. In good lighting conditions, three previous methods [Nguyen et al., 2011b][Bradley et al., 2007][Nguyen et al., 2012c] and the proposed one have very high detection rates. However, when different illumination effects are taken into account, the three previous methods degrade sharply while the proposed method still maintains reasonable performance. Overall our algorithm outperforms the others in different scenarios as well as under different illumination effects in outdoor environments. Image databases and videos related to this paper are available at [Nguyen, 2012].

Regarding implementation and running time issues, step 1 and 2 runs with $O(mlogm)$ time. The step 4 also requires an $O(mlogm)$ time. Step 5 takes $O(m)$ time. The repetition from step 3 over time until reaching a tolerance parameter needs $O(m^2 logm)$ time. Therefore, our algorithm runs with $O(m^2 logm)$ time. Concretely, for each frame (640x480 pixels), a computer with CPU Intel Core 2 Dual CPU L7500 2x1.67 GHz and 1G of RAM takes around 145 ms to 348 ms for processing.

6. SPREADING ALGORITHM FOR EFFICIENT VEGETATION DETECTION

Figure 6.2: The figure shows five examples of multi-spectral data and results. The first column contains original images. The second column shows near-infrared images. The third column illustrates vegetation detection results using the NDVI. The last column demonstrates vegetation detection results using the MNDVI.

6. Spreading Algorithm for Efficient Vegetation Detection

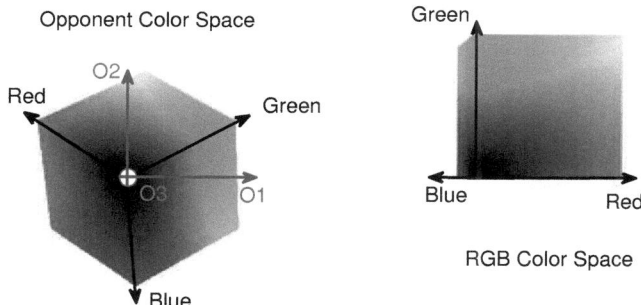

Figure 6.3: The opponent colour space (left) is obtained by rotating the RGB colour space (right) and swapping two channels R and G.

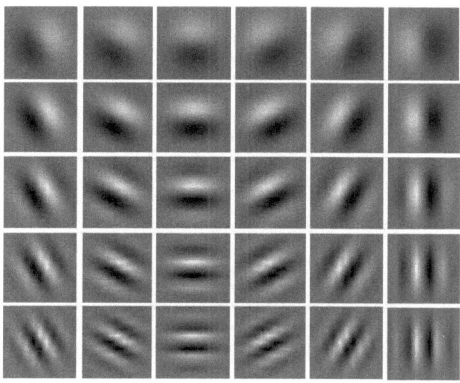

Figure 6.4: Gabor filter kernels in different scales in rows and orientations in columns.

6. SPREADING ALGORITHM FOR EFFICIENT VEGETATION DETECTION

Figure 6.5: From left to right: original image; segmented image; unstructured texture intensity; confidence map.

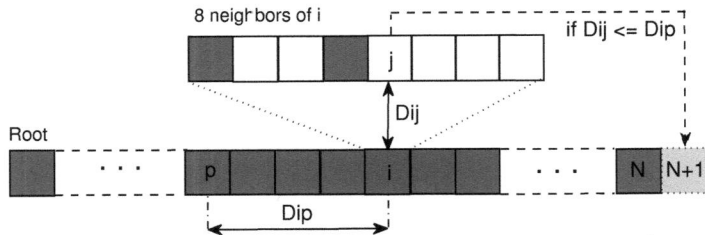

Figure 6.6: Vision-based spreading algorithm. Seed pixels are marked as dark green while the others are white. (For interpretation of the references to colour in this figure legend, the reader is referred to the electronic version of this dissertation.)

6. Spreading Algorithm for Efficient Vegetation Detection

Figure 6.7: From left to right: colour image; NIR image; spectral reflectance-based spreading mask; vision-based spreading mask.

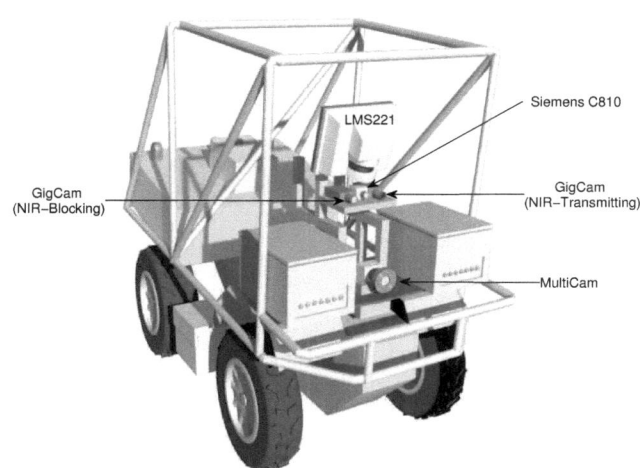

Figure 6.8: A model of our autonomous mobile outdoor robot.

6. SPREADING ALGORITHM FOR EFFICIENT VEGETATION DETECTION

Figure 6.9: The first row shows original images. Segmented images are illustrated in the second row. The third row shows the unstructured texture intensities. The fourth row presents the confidence maps. The last row demonstrates the results given by the algorithm.

6. Spreading Algorithm for Efficient Vegetation Detection

Table 6.1: Confusion Matrices of Different Approaches for Different Groups of Scenes

		Good Lighting		Under-Exposure		Over-Exposure		Shining		Shadow	
		Pos (%)	Neg (%)	Pos (%)	Neg (%)	Pos (%)	Neg (%)	Pos (%)	Neg (%)	Pos (%)	Neg (%)
Lalonde[a]	Pos	48.25	51.75	48.34	51.69	45.52	54.48	51.02	48.98	49.66	50.34
	Neg	41.11	58.89	42.07	57.93	40.36	59.64	47.23	52.76	48.24	51.76
Nguyen[b]	Pos	84.29	15.71	65.82	34.18	52.88	47.12	55.63	44.37	62.55	37.45
	Neg	30.54	69.46	36.41	63.59	40.53	59.47	35.14	64.86	36.02	63.98
Bradley[c]	Pos	94.62	5.38	77.66	22.34	75.68	24.32	65.51	30.49	88.28	11.72
	Neg	16.23	83.77	30.62	69.38	28.16	71.84	19.35	80.65	35.14	64.86
Nguyen[d]	Pos	92.70	7.30	71.25	28.75	85.16	14.84	78.39	22.61	86.33	13.67
	Neg	11.69	88.31	19.77	80.23	31.22	68.78	31.80	68.20	26.66	73.34
Proposed	Pos	95.45	4.55	90.21	9.79	87.49	12.51	85.01	14.99	90.18	9.82
	Neg	9.27	90.73	14.35	85.65	15.10	84.90	13.83	86.17	19.59	80.41

[a] Local Point Statistic Analysis, [Lalonde et al., 2006]
[b] 2D-3D Feature Fusion, Nguyen et al. [2011b]
[c] Vegetation Indices, Bradley et al. [2007]
[d] Modification of Normalized Difference Vegetation Index, Nguyen et al. [2012c]

6. SPREADING ALGORITHM FOR EFFICIENT VEGETATION DETECTION

6.6 Conclusion

We have introduced a spreading algorithm for vegetation detection using a multi-spectral approach. Compared with previous approaches, our algorithm provides a robust vegetation detection against illumination effects, which is both quantitatively more accurate and visually more pleasing. Instead of giving a specific model of colour or texture, our algorithm investigates discriminative characteristics of vegetation based on novel statements in properties 1,2,3,4. This enables the algorithm to detect varieties of vegetation that appear in different colours and textures. On the other hand, using a multi-spectral approach helps to catch the most discriminative characteristic of vegetation in light absorption and reflectance, which is distinguishable from other objects in general. Overall, the algorithm enables a fast and robust vegetation detection module which can be used for automobile navigation guidance when driving in complex environments. We acknowledge that the use of a Multi-Cam has some limitations regarding to low NIR image resolution (PMD sensor) and narrow sensor sensitivity (CMOS sensor). A desired vegetation system should have a multi-spectral system, which consists of two high dynamic range (HDR) cameras, and an active NIR lighting source. Such a system is planned to be established to test the performance of the proposed algorithm in a future work. Additionally, the use of polarizing filter, which helps to enhance the colour image with respect to reducing contrast, should be also taken into account in order to reduce the changes of MNDVI and NDVI with illumination changes.

Chapter 7

A Novel Approach for a Double-Check of Passable Vegetation Detection in Autonomous Ground Vehicles

The chapter introduces an active way to detect vegetation which is at front of the vehicle, in order to give a better decision-making in navigation. Blowing devices are to be used for creating strong wind to effect vegetation. Motion compensation and motion detection techniques are applied to detect foreground objects which are presumably judged as vegetation. The approach enables a double-check process for vegetation detection which was done by a multi-spectral approach, but more emphasizing on the purpose of passable vegetation detection. In all real world experiments we carried out, our approach yields a detection accuracy of over 98%. We furthermore illustrate how the active way can improve the autonomous navigation capabilities of autonomous ground vehicles.

This work has been published in **Proceeding of 15th IEEE Conference on Intelligent Transportation System (ITSC-2012)** [Nguyen *et al.*, 2012d].

7. PASSABLE VEGETATION DETECTION

7.1 Introduction

Regarding to the literature of robotics research, to increase autonomous ground vehicle (AGV) safety and efficiency on outdoor terrains, the vehicle's control system should have different strategies and settings for individual terrain surfaces. To enable more autonomous tasks in complex outdoor environments, the vehicle must have more "feeling" and "seeing" [Boley et al., 1989][Iagnemma & Dubowsky, 2002][Sadhukhan & Moore, 2003][Ojeda et al., 2006] [DuPont et al., 2005][Collins, 2008][Angelova et al., 2007][Halatci et al., 2007][Rankin & Matthies, 2008][DuPont et al., 2008]. While good terrain models and terrain classification techniques are already available to deal with a variety of terrain surfaces, the key limitation of outdoor autonomous navigation is to cope up with domains at which the vehicle has to navigate through tall grass, small bushes, or forested areas. Since, current perception systems can not do effective obstacle detection in these conditions, an idea to detect vegetation areas and try to set up a new definition of an obstacle as vegetation is really appreciated. Indeed, a lethal obstacle is conventionally defined as a solid object with significant height, which soon presents problems. In situations such as a cornfield, a field of thick and tall grass, there may be dense geometric obstacles on all sides of the robot. This can lead to the vehicle getting stuck. In contrast, the vehicle can try to drive over vegetation without any damage that enables more autonomous tasks in agricultural applications, rescue mission, or even military operations.

Recently, there was large amount of research investigating on vegetation detection based on vision techniques and LiDAR-based terrain models [Nguyen et al., 2011a][Nguyen et al., 2011b][Nguyen et al., 2010b][Lalonde et al., 2006][Macedo et al., 2000][Lu et al., 2009] [Wellington et al., 2006]. However, different species of vegetation have different colours, textures, structures as well as shapes. Also, illumination changes in outdoor environments cause a huge impact on the quality and reliability of the detection methods. These restrict the applicabilities of those approaches for the purpose of detecting vegetation in general.

Alternatively, vegetation needs sunlight to survive, using chlorophyll to convert radiant energy from the sun into organic energy. Chlorophyll exhibits unique ab-

7. Passable Vegetation Detection

sorption characteristics, absorbing wavelengths around the visible red band (645 m), while being transparent to wavelengths in the near-infrared (NIR) (700 m)[Ünsalan & Boyer, 2004]. These characteristics of chlorophyll are commonly used to design indices to estimate the local vegetation density in the satellite remote sensing field [Shull, 1929][Jordan, 1969][Rouse et al., 1974][Huete, 1988]. [Nguyen et al., 2012c][Nguyen et al., 2011c] and [Bradley et al., 2007] investigated this discriminative property of vegetation to apply for detecting vegetation in autonomous ground vehicles. However, those works remarked that on-board navigation reveals much more complication than in multi-spectral satellite or airborne, with presence of shadow, shining, under- and over-exposure effects. Whereby, light spectral reflectance of objects changes significantly against these effects, thus, a direct-applied vegetation index into robotics alone could not provide a trust-able result for safe navigation. Therefore, [Bradley et al., 2007] had to combine the vegetation indices with 3D-features given by laser data analysis for a double-check. [Nguyen et al., 2012c][Nguyen et al., 2011c] suggested to use an active lighting system to create more independence with different sunshine conditions. Even though the approaches based on vegetation indices perform high accuracy and efficiency, the question regarding to traversability is not yet answered.

In this context, we are going to answer the question of traversability by classifying vegetation into two classes: navigable and non-navigable. For that aim, we first try to figure out which vegetation can be passable for an AVG. For an easier understanding the case, let us start to discriminate between a stand of grass and a roll of barbed wire, or between cornstalks and thin trees. Respecting to the chlorophyll-light spectral synthesis, the more chlorophyll a material has, the easier it is to drive through. Grass and cornstalks contain richer chlorophyll, so they are easier to drive through. This property can be exploited using a multi-spectral approach. Particular in this work, we follow the works of [Nguyen et al., 2012c] and [Bradley et al., 2007]. On the other hand, regarding to kinematic consideration, grass and cornstalks are easy to drive through because of less resistance. In other words, grass and cornstalks are softer and movable, which can be clearly seen that they are easier to be moved under blowing wind. In order to utilize this characteristic, we suggest to use

7.2. MULTISPECTRAL-BASED VEGETATION DETECTION

an air compressor device to create strong wind. The movement of vegetation will be detected and recorded to set levels of "resistance". Overall, vegetation with rich chlorophyll and less resistance should be navigable one, therefrom comes the title double-check of passable vegetation detection. The structure of the paper is organized as follows: Section 7.2 describes how to index vegetation respecting to light spectral reflectance. Section 7.3 introduces the system design of our robot while an active way to measure the resistance of vegetation is illustrated in section 7.4. Experiments and results will be discussed in section 7.5 while section 7.6 concludes the work.

7.2 Multi-spectral-based Vegetation Detection

7.2.1 Standard Form of Vegetation Index

Similar to the intense reflection of fluorescent light from snow, vegetation reflects strongly in all direction the light in the near-infrared band. On the other hand, the photosynthesis process of chlorophyll inside vegetation requires more light spectral absorption in the red and blue bands. Let's recall the Normalized Difference Vegetation Index as mentioned in Chapter 3.

$$NDVI = \frac{\rho_{NIR} - \rho_{Red}}{\rho_{NIR} + \rho_{Red}} \qquad (7.1)$$

[Bradley et al., 2007] applied this index quite successfully in the field of robotics under good lighting conditions. When considering more illumination effects, the changes in light reflectance in the near-infrared and red bands are not linear, thus, NDVI can not be used efficiently to detect vegetation [Nguyen et al., 2012b]. Concretely NDVI of pigment metals, dark wet soil or black polymer synthesis materials in many circumstances might be even higher than of vegetation.

7.2. Multispectral-based Vegetation Detection

7.2.2 Modification Form of Vegetation Index

In order to be stable with respect to illumination effects in outdoor environments, [Nguyen et al., 2012c] proposed to use a MultiCam which integrates a CMOS sensor and Photo-Mixer Device (PMD) sensor into a molecular set-up [Ghobadi et al., 2010]. The MultiCam has its own infrared lighting system with the wavelengths centred at 870 nm. The intensity of the lighting source is adjustable, which lets a chance to stabilize the received NIR reflectance values. [Nguyen et al., 2012c] illustrated that there was a linear proportion of illumination to red but logarithm proportion to NIR. Thus, a better fit of normalized difference vegetation index was devised as follows [Nguyen et al., 2012c]

$$MNDVI = \frac{\rho_{NIR} - log(\rho_{Red})}{\rho_{NIR} + log(\rho_{Red})} \quad (7.2)$$

Our previous work [Nguyen et al., 2012c] has shown that MNDVI performs much better than NDVI in classifying vegetation and non-vegetation under different lighting conditions while taken into account shadow, shining, and overexposure effects.

7.2.3 Convex Combination of Vegetation Indices

The logarithmic term in **E.q 7.2** expresses the less impact of the red reflectance when an artificial lighting system is used. However, the softening red reflectance impact in MNDVI index is presenting problems in applied in an under-exposure or dim lighting condition where the logarithm term approaches to zero. In contrast, NDVI reveals good performance in these circumstances but failed to deal with strong shining and over-exposure effects. Therefore, in this work, we propose a convex combination of both the indices and supposed to be less sensible against illumination changes.

$$VI_{norm} = \alpha \times MNDVI + (1 - \alpha) \times NDVI \quad (7.3)$$

7. PASSABLE VEGETATION DETECTION

where,

$$\alpha = \begin{cases} 1, & \text{if } RED > T_{expo} \\ 0, & \text{otherwise} \end{cases}$$

T_{expo} is manually set to define the state of dim lighting or under-exposure on the red channel (in our case, $T_{expo} = 0.3$ when the red values are normalized). Thus, the NDVI index is only used in case of under-exposure or dim lighting condition, otherwise vegetation detection relies on the MNDVI index. **Fig. 7.1** illustrates examples of vegetation detection results based on NDVI, MNDVI and VI_{norm}, respectively. The results perform a good supplement between the two forms of vegetation indices against illumination changes. To have a more quantitative persuasion, we provide the confusion matrices of vegetation detection based on different vegetation indices as in **Table 7.1**[1]. Whereby, MNDVI and VI_{nortm} perform better than NDVI. The VI_{norm} index increases the true positive precision rate but also allows more false positive compared with the MNDVI. This issue can be covered when combined with the active method introduced in the next section.

Table 7.1: Confusion Matices of Different Vegetation Indices

	NDVI [Rouse et al., 1974]		MNDVI [Nguyen et al., 2012c]		VI_{norm} (this approach)	
	Pos	Neg	Pos	Neg	Pos	Neg
Pos	86.24	13.76	90.24	9.76	93.47	6.53
Neg	22.51	77.49	14.38	85.62	18.32	81.69

7.3 System Design

The main configuration in details of our autonomous mobile outdoor robot (AMOR) is described in [Kuhnert & Seemann, 2007][Kuhnert, 2008][Nguyen et al., 2012c][Nguyen et al., 2011c][Nguyen et al., 2011b]. As mentioned in the introduction part, for this

---

[1] *The evaluation was carried out with 500 outdoor images captured in both morning and afternoon conditions.*

7. Passable Vegetation Detection

Figure 7.1: Example of vegetation detection results given by different vegetation indices. The first column illustrates original images. The second column describes detection results given by the NDVI approach. The third column shows results of MNDVI approach. The last column demonstrates the results from VI_{norm} approach.

particular task, we need a blowing device to create wind to effect vegetation. One might immediately think about utilizing the available air compressor of the robot's air-break system. This, however, is not a reliable solution. The robot lasts his battery quickly because of high power consumption for the charging process of the air compressor. The blowing duration is very short due to the small air compressor tank. More seriously, using the air compressor would affect to the break system, thus, potentially causes an unexpected movement of the robot. Then, we come up with an idea of using independent blowing devices. Take a look at current products for such work, we find Bosch leaf blowers such as Bosch ALB 18 LI Cordless Li-Ion and Bosch ALS 25 which are really suited for the work and quite cheap, at around 80 Euro. Indeed, the leaf blowers can run continuously for 10 minutes at blow speeds of up to 215 km/h. Meanwhile, the robot only needs to turn on the blowing device in case of facing vegetation as an obstacle, and for each time the blowing duration

7. PASSABLE VEGETATION DETECTION

required is just from five to ten seconds. Therefore, after each fully charge, the device can be used for at least 60 halt states, which is so far satisfying us at the current stage.

Since the aim is to detect tall grass or branches of leaves which block the path of the robot, our interested area is basically located at the front. The goal is to cover the whole area with wind, so many blowing devices should be used. The number of the devices used depends heavily on the size of the robot to ensure that all front obstacles are effected by the wind created by those devices. Also, it would be a waste of money to have more than what is needed. In our case, the robot is 2.5 m long, 1.1 m wide, and 1.8 m high, so we need to use six blowing devices mounted at the middle and two sides of the robot (see **Fig. 7.2**), at a height around 85 cm. Practical experiments show that the distance of 30 cm from one device to its horizontal neighbour is reasonable for the wide cover. For the high cover, we only need to use other pipes with the heads bending down 30 degrees. It is not necessary to have similar ones with the heads bending up because the robot should not try to

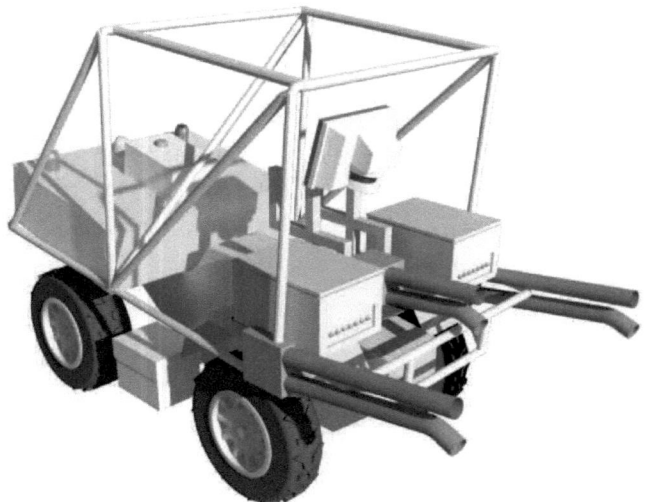

Figure 7.2: The AMOR model is shown here where six blowing devices corresponding with six pipes are mounted at front of the robot (figure provided by J. Schlemper).

drive over obstacles including vegetation with the height more than 1.2 m, which is out of our interest. The diameter of all pipes is 7 cm. The design really meets our aim for the tasks of driving over tall grass and passing though a narrow road with many branches of leaves bending down, which have been the main tasks in European Land Robot Trial (ELROB) since 2007.

7.4 A Double-Check for Passable Vegetation Detection

As a general rule, the richer chlorophyll the material has, the easier it is to drive through. Hence, vegetation revealing high values of VI_{norm} tends to be passable. A double-check of passable vegetation detection can be done by considering the resistance property of vegetation with respect to kinematic consideration. For that aim, we implemented an air compressor device to create strong wind to effect vegetation. Actually, the problem given to be solved in this work is that the vehicle gets stuck in a corn field or tall grass area or the path is blocked by a branch of leaves. Now, the vehicle has to decide which way provides less resistance based on detecting passable vegetation. So, in this application domain, the vehicle is in halt state and processing time is not extremely critical. Ideally, background subtraction techniques such as Mean and Covariance [Wren et al., 1997], Mixture of Gaussians [Grimson, 1998], Normalized Block Correlation [Matsuyama et al., 1999], Temporal Derivative [Haritaoglu et al., 1998], Bayesian Decision [Nakai, 1995], Eigen-Background [Oliver et al., 2000] and Wallflower [Toyama et al., 1999] could be directly used to establish a background model for the scene before winded. This leads to detect vegetation as foreground objects when blown by the air compressor device. However, even in the halt state, the vehicle has its own vibration created by the engine as operating, which degrades the quality of those background subtraction techniques. In order to assure a robust motion detection, a motion compensation process is necessary. The following explanation reveals how to compensate the vibration of the vehicle. First, with the high speed of frame rate from the MultiCam, at about 30 fps, we can approximately

7. PASSABLE VEGETATION DETECTION

assume that the movement of objects in the scene is rather slow and has a brightness constancy. Assume that after the small time δt, the frame is shifted $(\delta x, \delta y)$ and the rotation is θ. If (x', y') is the next stage of (x, y) then,

$$\begin{cases} x' = (x+\delta x)\cos\theta + (y+\delta y)\sin\theta \\ y' = -(x+\delta x)\sin\theta + (y+\delta y)\cos\theta \end{cases}$$

with the assumption of small movement, we have

$$\begin{cases} x' \approx x + \delta x + y\theta \\ y' \approx -x\theta + y + \delta y \end{cases}$$

Assume a brightness constancy, so

$$I(x', y', t+\delta t) = I(x, y, t) \tag{7.4}$$

where, $I(x,y,t)$ is the intensity value of a point at the position (x,y) of the frame taken at the time t. $I(x',y',t+\delta t)$ is the point when moved to the position (x',y').

From Taylor Series Expansions,

$$I(x',y',t+\delta t) \approx I(x+\delta x+y\theta, -x\theta+y+\delta y, t+\delta t)$$
$$\approx I(x,y,t+\delta t) + \frac{\partial I}{\partial x}\delta x + \frac{\partial I}{\partial y}\delta y + \frac{\partial I}{\partial x}y\theta - \frac{\partial I}{\partial y}x\theta$$

substitution into **Eq. 7.4**,

$$I(x',y',t+\delta t) - I(x,y,t) \approx I(x,y,t+\delta t) - I(x,y,t) + \frac{\partial I}{\partial x}\delta x + \frac{\partial I}{\partial y}\delta y + \frac{\partial I}{\partial x}y\theta - \frac{\partial I}{\partial y}x\theta \approx 0$$

7. Passable Vegetation Detection

or the difference of two adjacent frames can be written as,

$$I(x,y,t+\delta t) - I(x,y,t) = \frac{\partial I}{\partial x}\delta x + \frac{\partial I}{\partial y}\delta y + \frac{\partial I}{\partial x}y\theta - \frac{\partial I}{\partial y}x\theta \approx 0 \qquad (7.5)$$

Solving **E.q 7.5** throughout the images using least square fit algorithm, the returned parameters of displacement and rotation help to compensate the vibration of the vehicle. The interesting point is that when the robot stops on a slope, its vibration might be significant whereby the small movement assumption is no more valid. In this case, the estimation given by the above vision-based method is not usable. Therefore, we propose to use the Inertial Measurement Unit (IMU) information as a reference to judge whether the above vision-based motion estimation is trustable. In an untrustable case, the IMU information is used instead of the estimated parameters given by the vision-based method.

Figure 7.3: The first column describes original image and vegetation detection by by VI_{norm}. The second column shows accumulative background subtraction using Mean & Threshold without and with motion compensation, respectively. The last row illustrates accumulative background subtraction using Mixture of Gaussians without and with motion compensation, respectively.

7. PASSABLE VEGETATION DETECTION

In general, the proposed motion compensation algorithm helps to have a better background subtraction, which has been proved by applying two background subtraction techniques with and without the motion compensation process as demonstrated in **Fig. 7.3**. A performance comparison between all current background subtraction techniques with and without the motion compensation process might be interesting, we, however, have not yet done it due to our satisfaction with the background subtraction result given by the Mixture of Gaussians.

The problem now is that even if we have detected the movement areas in the scene, how should we know which parts are most likely moved significantly? Then, we decide to record all movements of moving pixels in the scene. Look into a local region of the current frame at position (x,y) or look back to **Eq. 7.4**. We apply another Taylor Series Expansions as follows,

$$I(x',y',t+\delta t) \approx I(x+\delta x+y\theta, -x\theta+y+\delta y, t+\delta t)$$
$$\approx I(x,y,\theta,t) + \frac{\partial I}{\partial x}\delta x + \frac{\partial I}{\partial y}\delta y + \frac{\partial I}{\partial x}y\theta - \frac{\partial I}{\partial y}x\theta + \frac{\partial I}{\partial t}\delta t$$

hence,

$$\frac{\partial I}{\partial x}\delta x + \frac{\partial I}{\partial y}\delta y + \frac{\partial I}{\partial x}y\theta - \frac{\partial I}{\partial y}x\theta + \frac{\partial I}{\partial t}\delta t \approx 0 \qquad (7.6)$$

When already compensated, there should be no consideration in rotation but only local movement of vegetation, so $\theta = 0$. **Eq. 7.6** is re-written as,

$$\frac{\partial I}{\partial x}\frac{\delta x}{\delta t} + \frac{\partial I}{\partial y}\frac{\delta y}{\delta t} + \frac{\partial I}{\partial t} \approx 0 \qquad (7.7)$$

Or,

$$I_x V_x + I_y V_y \approx -I_t \qquad (7.8)$$

with I_x, I_y are the derivatives, $V_x = \delta x/\delta t$ is the velocity in the horizontal axis, $V_y = \delta y/\delta t$ is the velocity in the vertical axis.

Eq. **7.8** is re-written as

$$I_\Delta^T V = -I_t \qquad (7.9)$$

with $I_\Delta^T = [I_x\ I_y]$ and $V^T = [V_x\ V_y]$. **Eq. 7.9** is such a familiar equation expressing the relationship between velocities and derivatives in optical flow problems. The main idea is that the movement of living vegetation is most likely a damped oscillation after a blowing process given by the air compressor device. Therefore, the block diagram of our algorithm is sketched as in **Fig. 7.4**. Assume that the vehicle captured M frames before and N frames after the blowing process. Background subtraction is carried out and accumulated to result the final accumulative background subtraction, thus, the movement of vegetation should lie in the part marked as foreground (see **Fig. 7.3**). The masks of the accumulative foreground (MAF) extracted from the final accumulative background subtraction and of detected vegetation (MDV) from VI_{norm} are merged to generate the mask of possible dynamic vegetation(MPDV). The movement of every pixel in the MPDV is recorded by the optical flow process to weight the resistance of those vegetation pixels. Many optical flow algorithms can be applied to record the movements of foreground objects such as in [Lucas & Kanade, 1981], [Horn & Schunk, 1981], [Farnebäck, 2003], and [Brox et al., 2009]. Regarding to the particular case of passable vegetation detection, some challenging issues often found in the optical flow problem such as *aperture problem [Ullman, 1979], sudden lighting change [Toyama et al., 1999]* are approximately not influential, thus, we propose to use a simple method like dense optical flow [Farnebäck, 2003] to do the work. Also, the work [Farnebäck, 2003] demonstrated that the dense optical flow technique shows out-performance compared with others when taken into account both computation and precision for a two-frame algorithm.

7.5 Experiments and Results

We used an autonomous ground vehicle with the configuration in details described in [Nguyen et al., 2012c][Nguyen et al., 2011c][Nguyen et al., 2011b] and **Fig. 7.2**

7. PASSABLE VEGETATION DETECTION

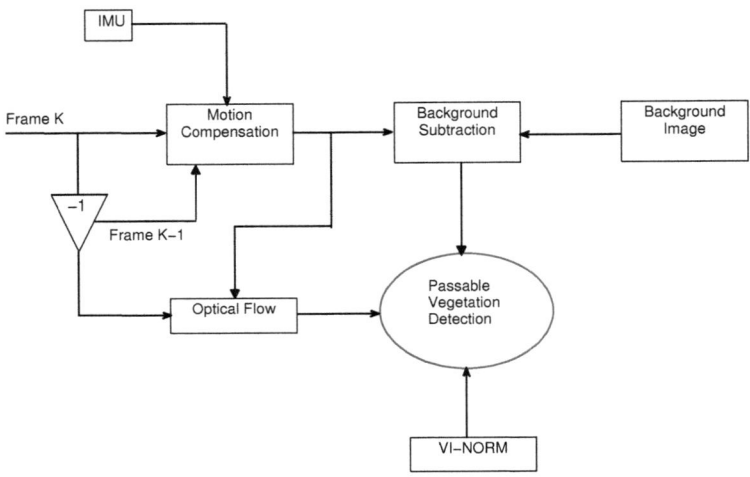

Figure 7.4: Block Diagram of the Proposed Algorithm.

for an evaluation of the proposed algorithm. All data was collected and stored in the robot's computer when the robot traversed throughout outdoor environments in both morning and afternoon conditions. Colour images were firstly segmented into small regions with respect to homogeneous colour [Felzenszwalb & Huttenlocher, 2004]. Whereby vegetation regions were hand-labelled as ground truth to evaluate the outputs of the algorithm.

Table 7.2: Confusion Matix of Passable Vegetation Detection

	Passable Vegetation	others
Passable Vegetation	98.76	1.24
others	2.01	97.99

The quantitative evaluation shown in **Table 7.2** is carried out with 1000 input images captured from 50 halt states of the vehicle (200 frames per each halt state). The result is quite convincing with high accuracy of detecting and weighting passable vegetation, at about 98.37%. Notice that in this work, the vegetation with intense movement after the blowing process is determined as passable vegetation. Thus,

7. Passable Vegetation Detection

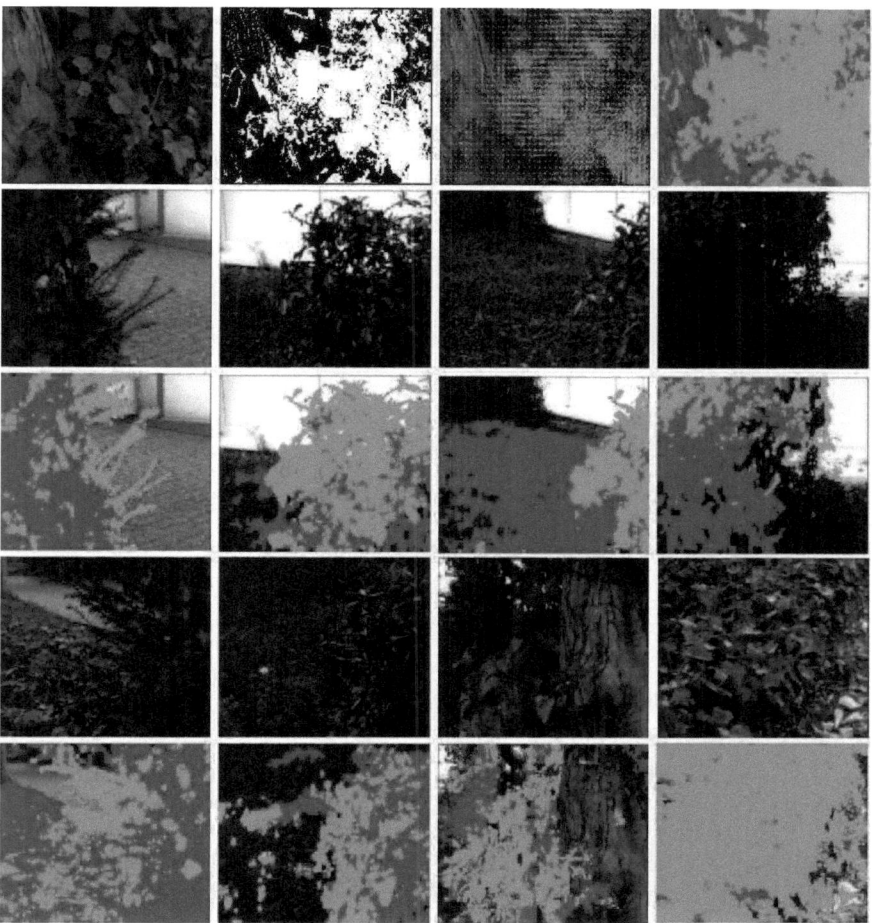

Figure 7.5: The first row, from left to right, illustrates original, background subtraction, optical flow and result images, respectively. The second row and fourth row show original images while the third row and the fifth row describe the outputs from our algorithm, respectively. The green and dark green colours reveal passable and non-passable vegetation detected in the result images, respectively.

vegetation which is not effected by the blowing wind due to far distance to the vehicle or out of the wind flow is detected as non-passable vegetation, which would not

7. PASSABLE VEGETATION DETECTION

be taken into account for evaluating passable vegetation detection. In other words, passable vegetation detection accuracy is only evaluated inside the area effected by the blowing wind. Alternatively, examples of passable vegetation detection are illustrated in **Fig. 7.5** to have a better intuitive demonstration. Indeed, we can clearly see that branches of leaves and vegetable are successfully detected as passable vegetation, which will be then utilized to enhance decision-making in navigation. One with good observation might recognize that low grass is detected as non-navigable vegetation (marked with dark green) because the movement of low grass is much lesser than of leaves, which is usually confused as the vibration of the robot. It is infeasible to distinguish between the small movement of low grass and the small movement caused by the vibration of the robot even motion compensation already done. However, this issue can be simply resolved by taking the height information into account. Thus, vegetation with low height or less-resistance should be navigable one.

7.6 Conclusions

We have introduced an active way for a double-check of passable vegetation detection, which helps to have a better decision-making in outdoor navigation especially in complex outdoor environments with the presence of dense vegetation. Unlike previous approaches in vegetation detection, the proposed approach is not to be significantly affected by visual effects or illumination changes. A double-check between a multi-spectral and an active approaches leads to a more realistic and efficient mechanism for detecting vegetation respective to the purpose of classifying navigable or non-navigable ones. The approach has been implemented and evaluated in several real-world experiments. The experiments show that our approach is able to accurately detect tall or low grass and branches of leaves with an accuracy of more than 98%. The current approach is limited to detecting and weighting passable vegetation in a halt state of the vehicle. In a future work, we will investigate whether the described approach can also be applied in case of a running vehicle.

Chapter 8

Terrain Classification Based on Structure for Autonomous Navigation in Complex Environments

One of the main challenges for autonomous navigation in cluttered outdoor environments is to determine which obstacles can be driven over and which need to be avoided. Especially in off-road driving, the aim is not only to recognize the lethal obstacles on the vehicle's way at all costs, but also to predict the scene category thereby giving a better decision-making framework for vehicle navigation. This chapter studies terrain classification based on structure, which relies on sparse 3D data from LiDAR mobility sensors. While most of recent methods for LiDAR data processing are purely found on the local point density and spatial distribution of the 3D point cloud directly. We, on the other hand, introduce a new approach to analyse the point cloud by considering local properties and distance variation of pixels inside edgeless areas. First of all, the edgeless areas are extracted from segmenting the 3D point cloud into homogeneous regions by Efficient Graph-based technique. Secondly, the neighbour distance variation inside edgeless regions (NDVIER) features are obtained

8. TERRAIN CLASSIFICATION BASED ON STRUCTURE

by calculating the euclidean distance of neighbour distance variation inside each region. Through extensive experiments, we demonstrate that this feature has properties complementary to the conditional local point statistics features traditionally used for point cloud analysis, and show significant improvement in classification performance for tasks relevant to outdoor navigation.

This work has been published in **Proceedings of International Conference on Communications and Electronics** [Nguyen *et al.*, 2010b]

8.1 Introduction

Recently, autonomous navigation techniques work well for environments such as hallways and on roads, where obstacles are static and usually rigid. The remaining problem is the difficulty in describing the environment of the vehicle in a way that captures the variability of natural environments. In order to solve the problem, there are many approaches proposed using different sensor systems, such as LiDAR, PMD camera and multi-spectral camera.

The multi-spectral camera simply produces multiple images in different spectral range for analysis. The interesting point is that a simple pixel-by-pixel comparison between red and near infrared ray (NIR) reflectance, normally referred to as a vegetation index, potentially provides a powerful and robust way to detect vegetation [Bradley *et al.*, 2007][Willstatter & Stoll, 1913]. However, the dependency of trusted data acquisition on the change of light intensity makes the camera's applicability unstable under the presence of the sky, or shadowed areas.

PMD camera is a real time active 3D range camera based on time-of light technology using the Photonic-Mixer-Device(PMD)[Plaue, 2006], which produces low-resolution images(64x48 pixels) of the depth and modulation amplitude at high frame rate, at around 50 to 60 fps, which is comparable to a regular camera. Therefore the camera is well suited for real-time object detection whereby interactive 3D data can be obtained by moving the camera around in space. Nevertheless, the camera does not operate properly in an outdoor environment because of the strong noise arising

8. Terrain Classification Based on Structure

with the presence of intense sunlight, smooth surfaces, and metals [Nguyen et al., 2010a].

On the other hand, the use of LiDAR has been proposed to get the stable depth data in outdoor environments [Tuley et al., 2004][Anguelov et al., 2005][Huang et al., 2000][Rasmussen, 2002]. A quite successful work to segment 3D point cloud into three classes: surfaces(ground surface, rocks, large tree trunk), linear structures(wires thin branches, small tree trunk) and porous volumes(foliage grass) is given by [Lalonde et al., 2006]. This technique is based on local spatial statistics extracted over a fixed-size support volume, so it is computationally expensive and also highly depends on the size of the applied window. In addition, the edge effect [Nguyen et al., 2010b] [1] affects to the linear property of extracted features, which usually causes confusion between edge points in a linear structure or solid surface and scattered points.

In this chapter, we address a simple but very efficient approach for 3D point cloud processing based on geometric structure to support ground vehicle mobility. We use the SICK laser LMS221 mounted front of the vehicle (see **Fig. 2.4** in section 2.1), so that, the 3D point cloud data are obtained by sweeping the laser vertically, see **Fig. 8.1**. The basic idea is that the artificial constructions, tree trunks, or roads are normally represented by linear structures or smooth surfaces while the vegetation representation is considered as a highly textured region.

In fact, there are several previous works having attempted to capture spatial texture analysis in order to facilitate image segmentation and interpretation. Still, the unsolved problem of edge effect lowers their complexity and makes them differ from the idea of being really robust. In order to totally remove the bad effect of edge reflectance, we, at first, segment the 3D point cloud into homogeneous areas by Efficient Graph-based technique. The distance variation of each pixel to its neighbours is calculated by the euclidean distance of neighbours' distances inside each homogeneous region. The result is considered as neighbour distance variance inside edgeless region (NDVIER) features which are quite discriminative where vegetation

---
[1] The edge effect here is understood as the scattering property of the edge points between two or more regions with big distance difference, which makes the edge points behaviour like scattered points do regarding spatial distribution.

8. TERRAIN CLASSIFICATION BASED ON STRUCTURE

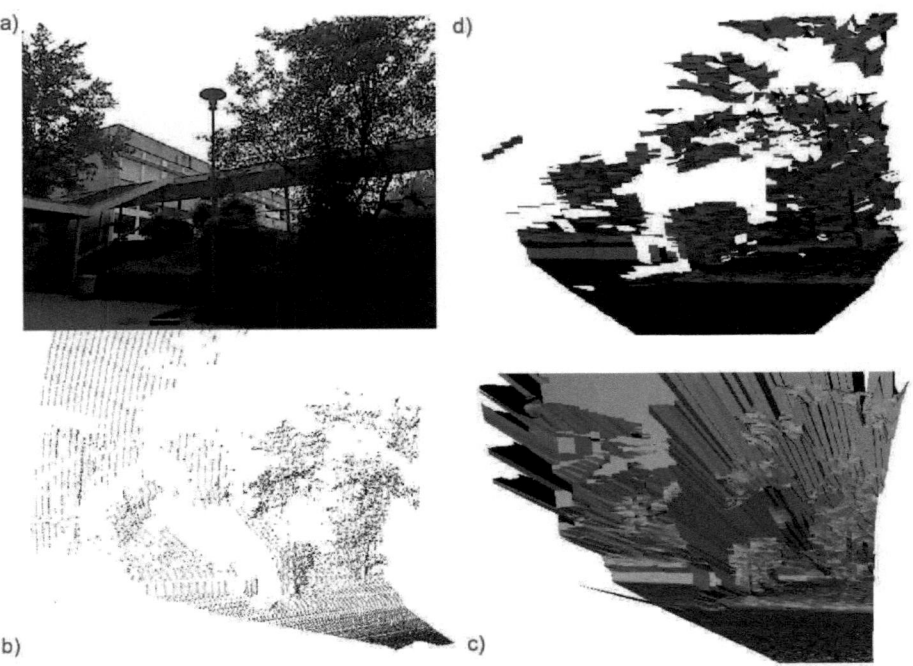

Figure 8.1: An example of 3D point cloud given by SICK LMS221 where a) colour image of the scene; b) 3D points in Cartesian coordinate (the maximum distance set is 16 meters, so all farther objects which are not in the case of consideration are illustrated by vertical lines with distance of 18 meters); c) Point cloud triangulation; d) 3D reconstruction of the scene with invalid faces removed.

areas perform high distance variation compared with others'. In order to achieve a more robust and complex detection, the 3D point distribution feature is taken into account with the main goal is to classify tree trunk and solid surface or roads, while the elevation information of each pixel is accounted to discriminate grass, bushes and leaves.

The chapter is structured as following: Section 8.2 explains the methodology of our approach to terrain classification based on structure of 3D point cloud given by a LiDAR. To give a demonstration of the proposed classification technique, section

8.3 discusses some experiments and results, while section 8.4 concludes this work.

8.2 Methodology

The traditional way for analysing 3D data given by a LiDAR is to capture the spatial distribution of points in local neighbourhood [Vandapel et al., 2004]. According to the state-of-the-art work at Carnegie Mellon University [Lalonde et al., 2006], the local spatial point distribution, over some neighbouring area, is captured by the decomposition into principal components of the covariance matrix of the 3D points position, ordered by decreasing eigenvalues. Intuitively, in the case of scattered points, there is no dominant direction between the points, so the eigenvalues are nearly equal to each other. In the case of linear structure, there should be only one dominant direction, so the first eigenvalue is much superior than the others. Finally, in the case of solid surface, the principle direction is aligned with the surface normal with the first two eigenvalues are close to each other and far different from others. The work of [Lalonde et al., 2006] demonstrated that these properties can be potentially used in describing outdoor environments. However, the approach of purely using the point distribution is not really robust in some scenes, especially with the presence of dense edges. In the case, the linear structure points are usually confused as the scattered points. Therefore, the task to overcome this issue is to eliminate edge points, as a result, we naturally come up with the idea of segmenting the 3D point cloud into homogeneous regions. K-mean technique, a quite popular classification technique in data analysis, is first used to do this task. Nevertheless, K-mean classifies data only relying on their values without concerning about the spatial distribution and local properties. This causes many drawbacks in recognizing or grouping real homogeneous regions. Fortunately, Efficient Graph-Based Image Segmentation given by [Felzenszwalb & Huttenlocher, 2004] has provided a very efficient technique in image segmentation. In this paper, we will demonstrate that the technique is also very efficient in segmenting the 3D point cloud regarding to the spatial distribution, local and global properties of points. The next part will describe the Efficient Graph-based

8.2. METHODOLOGY

technique briefly.

8.2.1 Efficient Graph-based Segmentation Technique

The idea of this technique is based on selecting edges from graph, where each pixel corresponds to a node in the graph, and certain neighbouring pixels (usually four neighbours) are connected by undirected edges. Weights on each edge measure the dissimilarity between pixels. Following is a brief explanation of how to use the Grap-based technique:

The input is a graph $G = (I, E)$, with N vertices and M edges. If the kth pixel and the jth pixel are neighbours and denote the vertices connected by the ith edge in the ordering, the dissimilarity between them is calculated as.

$$E[i] = |I[j] - I[k]| \tag{8.1}$$

Note: if I is a colour image, the dissimilarity should be combined of three colour channels' differences.

$$E[i] = |I[j]_r - I[k]_r| + |I[j]_g - I[k]_g| + |I[j]_b - I[k]_b| \tag{8.2}$$

In this paper, I is the depth image obtained from distance information given by the laser sensor. E is considered as an edge vector of size M (= 4N in case of considering four neighbours). Assume the kth pixel belongs to the region a, while j^{th} pixel belongs to the region b. If a and b are disjoint components and the dissimilarity between them is small compared to the internal difference of both those components, then merge the two components, otherwise do nothing. To give more power to users, the segmentation technique allows to input the minimum size of a region segmented. Hence, if the a and b are disjoint and the size of a or b is smaller than the minimum size, then merge a and b. Or:

Loop{ # Start joining vertices
   if {$(a! = b)$ && $(E[i] \leq Int[a])$ && $(E[i] \leq Int[b])$} **then**
      Merge(a, b);

8.2. Methodology

    Int[a,b] = E[i] + THRESHOLD$(size(a) + size(b))$;
else
    do nothing
end if
} # End of the process
# Start to merge small regions
if $\{(size(a) < minSize) \&\& (size(b) < minSize)\}$ **then**
    Merge(a, b);
    Int[a,b] = max(Int[a], Int[b]) + THRESHOLD$(size(a) + size(b))$;
end if

$E[i]$ is the dissimilarity between the kth pixel and the jth pixel, see **Eq. 8.1**. $Int[a]$ is the internal difference of the region a. The additional term $THRESHOLD(size(x)) = c/size(x)$, with c: constant number given by users, is applied to keep the global properties. For the range of number of scanned points from a LiDAR usually from 10000 to 100000, the value of c should be set at about 150 and the minimum size is at about 100 pixels. Initially, all vertices are considered as edge points, where each point is considered as an "initial region" or one component, and the values of the internal difference of all components are set equal to $THRESHOLD(1)$.

The result in this method, even made by greedy decision, is verified that it is neither too coarse nor too fine [Felzenszwalb & Huttenlocher, 2004]. Intuitively, **Fig. 8.2** and **Fig. 8.3** shows that the 3D point cloud is well segmented into regions regarding their distances and local properties.

8.2.2 Feature Extraction

8.2.2.1 Neighbour Distance Variation Inside Edgeless Regions

Definition 1: An edgeless region is defined as a region where there is no big distance difference between any set of subregions.

From this definition, all objects in the viewed scene exist in form of edgeless regions after segmented by the method in the subsection 8.2.1. Following describes

8.2. METHODOLOGY

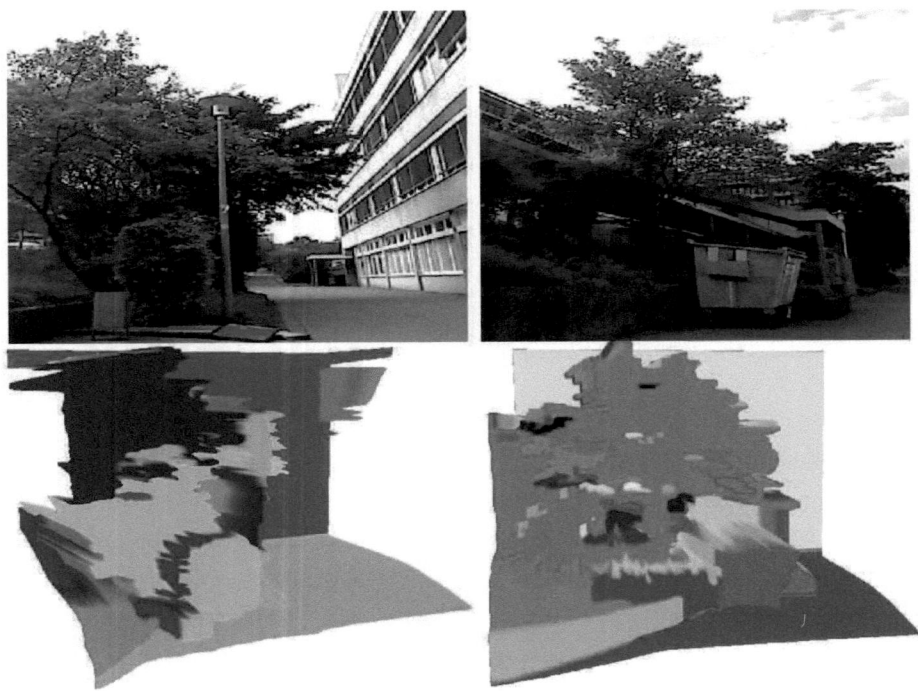

Figure 8.2: The first row shows colour images of the viewed scenes. The second row illustrates the corresponding results from point cloud segmentation (best viewed in colours).

how to calculate the neighbour distance variation of edgeless regions: the input data is a graph G = (V,R) where V is a matrix of 3D points position: $V_i = \{x_i, y_i, z_i\}$, R contains n regions of the depth image segmented in the previous subsection: $R = \{R_k; k = 1 : n\}$. We assume that the ith pixel belongs to the region R_k. We search all neighbour pixels which also belong to the region R_k. So, the neighbour distance variation of the ith pixel is calculated as:

$$NDVIER[i] = \sum_{j=0}^{m} \frac{|x[i] - x[j]| + |y[i] - y[j]| + |z[i] - z[j]|}{m \times dst[i]} \qquad (8.3)$$

8.2. Methodology

Figure 8.3: The first row shows colour images of the viewed scenes. The second row illustrates the corresponding results from point cloud segmentation (best viewed in colours).

$$\text{subject to } i = \{1:N\}.$$

where N: the number of 3D points.

m: the number of neighbour pixels which belong to the same region, $0 < m \leq 8$ (each pixel has maximum eight neighbour pixels).

A greedy decision is made by comparing the value of NDVIER[i] and the trained NDVIER values. NDVIER feature is scale invariant because it normalizes the neighbour distance variation and also takes into account the distance information of applied points to avoid the scattering effect of laser beam.

Property 1: Qualitatively, the high value of NDVIER refers the scatter structure

8.2. METHODOLOGY

(vegetation) of the selected volume, while the quite low value of NDVIER implies the solid surface.

Although, the intermediate value of NDVIER can not help to give any decision, it can be complemented with the conditional local point statistics feature (see the next subsection).

8.2.2.2 Conditional Local Point Statistics

These saliency features are inspired by the local point statistics approach of [Lalonde et al., 2006]. Instead of estimating directly 3D point distribution, the condition of the same region between input points has been proposed to avoid the effect of edge reflectance. From subsection A, the 3D point cloud is segmented into n homogeneous regions: $\{R_k, k = \{1 : n\}\}$. Assume that there is a set of M 3D points: $\{I_i\} = \{(x_i, y_i, z_i)^T\}$ with $i = \{1 : M\}$, in the same region R_k. The symmetric positive definite covariance matrix of the set is expressed as

$$Cov = \frac{1}{M} \sum_{i=1}^{M} (I_i - \bar{I})(I_i - \bar{I})^T \qquad (8.4)$$

With: $\bar{I} = \frac{1}{M} \sum_{i=1}^{M} I_i$. Actually, the raw data from LiDAR consists of information about vertical, horizontal angles, and distances. The values of x, y, z are easily computed by mapping from the Spherical coordinate to the Cartesian coordinate.

The principle components of the matrix are extracted, and named as eigenvectors: $\vec{e_0}, \vec{e_1}, \vec{e_2}$, and eigenvalues: $\lambda_0, \lambda_1, \lambda_2$, ordered by decreasing: $\lambda_2 \leq \lambda_1 \leq \lambda_0$.

Property 2: The relation of the three eigenvalues is normally referred as a spatial structure index where $\lambda_0 \approx \lambda_1 \approx \lambda_2$ denotes for scattering, $\lambda_0 >> \lambda_1 \approx \lambda_2$ denotes for linear structure, and finally $\lambda_0 \approx \lambda_1 >> \lambda_2$ denotes for solid surface.

Two saliency features can be obtained based on the Property 2: $S_{scatter} = \lambda_0$ and $S_{surface} = \lambda_1 - \lambda_2$; the so called *scatterness* and *surfaceness*, respectively. Regarding **Property 2**, the linear structure can be specified through $S_{linear} = \lambda_0 - \lambda_1$, but the practical proves that, the classification between linear structure and scattered is quite poor, at about 48% [Nguyen et al., 2010b]. Additionally, the task of detecting

8.2. Methodology

vegetation does not force to do such classification, so we do not use the linear feature in this case. In practice, the selection of M nearest points is also considered to a compromise between computational efficiency, memory management, and scene reconstruction accuracy. In the work of [Lalonde et al., 2006], the M nearest points were selected by sliding a cube of 10 cm edge across the 3D point cloud in space. The size of the cube was chosen experimentally in order to select the nearest points in the same object rather than in two or more different objects, which avoids the edge effect. This, however, seems subjective and highly depends on the type of laser sensor. In our approach, the segmentation of 3D point cloud, and then the selection of nearest points in the same segmented region, have implicitly done the work. Then, we suggest the mapping from the matrix of 3D point cloud into an array where every four neighbours are arranged continuously. The selection of M nearest points of 3D point cloud is now converted to the selection of a shifting interval of the array. **Fig. 8.4** shows the mapping between the 3D point cloud to the array of neighbour pixels with M=4.

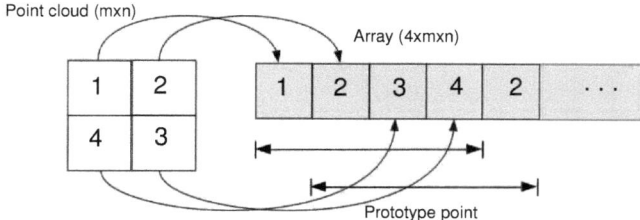

Figure 8.4: Mapping from 3D point cloud to an array of neighbour pixels. The selection of M (= 4) neighbours pixels in the 3D point cloud is actually taken place by capturing an interval of four numbers in the array, so called prototype point. The new prototype point is one pixel shift of the previous one.

8.2.3 Support Vector Machine

As described in the previous subsection that there are several parameters need to be set such as eigenvalues of the covariance matrix *Cov* and NDVIER values in order

8.2. METHODOLOGY

to clarify the **Property 1** and **Property 2** quantitatively. It is practically infeasible to hand-tune thresholds directly to result classification because those values highly depends on the type of environments, the type of sensors, the number of scanned points, and the point density. Experimental research shows that the variability of the values is manifested especially with the presence of tall grass (or dense edge areas), so we usually face with a nonlinear classification problem in the case of cluttered environments. In fact, there are many nonlinear classification techniques proposed at recent time, both supervised and unsupervised. While the supervised techniques usually cost computational expensive, the unsupervised ones are not well adapted to the nonlinear problems in reality. In this work, we train data with only four features in order to give the decision, so the supervised classification technique is preferred. We had tried to use Support Vector Machine (SVM) [Cortes & Vapnik, 1995], Naive Bayes classifier [Quinlan, 1993], Neuron Networks [Zhang, 2000], Adaboost [Freund & Schpire, 1997], and Expectation Maximization [Bilmes, 1997]. Consequently, proposed by Cortes and Vapnik in 1995, SVM shows out performance and is more reliable than others. Following is a brief description about SVM.

We, for instance, have L training samples, where each input x_i has D features and is in one of two classes $y_i = -1$ or $+1$. Assume that $y_i = +1$ denotes pedestrian samples(or positive samples) and $y_i = -1$ denotes non-pedestrian samples(or negative samples).

Inputs: $\{x_i, y_i\}$ where $i=1,2..L$, $y_i \in \{-1,1\}$, $x \in \Re^D$

The main task in training SVMs is to solve the following quadratic optimization problem:

$$min_\alpha f(\alpha) = \frac{1}{2}\alpha^T Q\alpha - e^T \alpha \ \ s.t \ \ 0 \leq \alpha_i \leq C, \ y^T \alpha = 0 \qquad (8.5)$$

Where e is the vector of all ones, C is the upper bound of all variables, Q is an L by L symmetric matrix with $Q_{ij} = y_i y_j K(x_i, x_j)$, and $K(x_i, x_j)$ is the kernel function. The kernel function is used because there are some classification problems that are not linearly separable in the space of the inputs x, which might be in a higher dimensionality feature space given a proper mapping $x \rightarrow \phi(x)$.

In our work, the training samples are a set of NDVIER, *scatterness, linearness* and *surfaceness* features with their corresponding hand-labelled classes. In order to achieve a more complex and accurate classification, the Kernel-Trick used is Radial Basic Function [Baudat & Anouar, 2001].

$$K(x_i, y_i) = exp(-\frac{||x_i - y_j||^2}{2\delta^2}) \qquad (8.6)$$

A cross validation process returns C = 75 with the stopping tolerance is set $\varepsilon = 0.01$.

8.3 Experiments and Results

In this work, 300 different scenes of cluttered outdoor environments are captured by the SICK laser LMS221 with 81x330 = 26730 pixels resolution and the maximum distance set is 16 m. The angular separation between laser beams is 1/4 degree over a 90^0 field of view. The angular separation between laser sweeps is 2/3 of a degree over 120^0. 200 3D point clouds are used for training and the other 100 are used for testing. The classification results are evaluated by comparing the output of the classifier with the hand-labelled data. In this paper, we evaluate the discrimination between scatter, linear, and surface structures rather than the specific classes of classification such as grass, trees, bushes, building, roads, etc. Actually, if we can have a good classification of the three structures, the object classification can lately be realized by evaluating the relationship between the object structure and the three structures. For example, the grass should be a vegetation area with little presence of linear structure, while the bushes and trees should be vegetation areas with dense presence of linear structure. The discrimination between trees and bushes can be clarified by estimating the elevation of their centroids. The roads and lethal obstacles can also be classified by their elevation regarding the discrimination of solid surface areas.

The classification processing time of our approach is short, at around 310 ms, however the acquisition time of the LiDAR is quite slow, at around 820 ms. Therefore, the total processing time of this approach is at around 1130 ms, which is not really reliable for on-board navigation. The main use of this approach is to detect

8. TERRAIN CLASSIFICATION BASED ON STRUCTURE

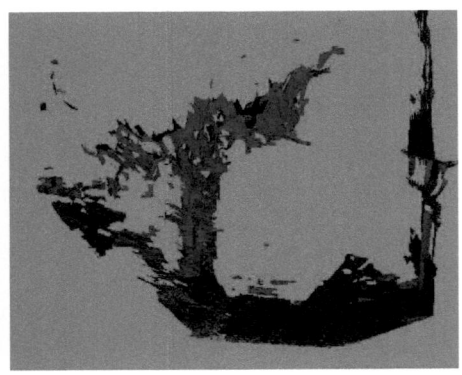

Figure 8.5: An example of 3D reconstruction of a 3D point cloud delivered by the SICK laser LMS221. The scene consists of flat area, grass, tree and wall.

Figure 8.6: An example of 3D reconstruction of a 3D point cloud delivered by SICK laser LMS221. The scene consists of building (at right hand), tree and flat area.

Figure 8.7: Example of data post-processing for the 3D point cloud in **Fig. 8.5**. The green colour denotes for vegetation areas, the dark blue colour denotes for linear structure areas, and finally the violet colour denotes for solid surface areas

Figure 8.8: Example of data post-processing for the 3D point cloud in **Fig. 8.6**. The green colour denotes for vegetation areas, the dark blue colour denotes for linear structure areas, and finally the dark cyan colour denotes for solid surface areas

roads or obstacles and predict the scene category front of the vehicle. **Fig. 8.5**, **Fig. 8.6** and **Fig. 8.7**, **Fig. 8.8** are the 3D reconstruction images from a 3D point cloud and the corresponding results given by our approach, respectively.

Table 8.1: Classification accuracy

Confusion Matrix (%)	scatter	linear	surface
scatter	58.2	33.1	8.7
linear	20.4	68.8	11.8
surface	2.3	12.6	85.1

Table 8.1 shows the classification accuracy of the results, which is 10 % to 17 % better than previous approaches' which purely rely on 3D point distribution. The discrimination between surface and scatter points is quite successful from our approach, while the linear structure and the solid surface, however, are more confused because of spurious misclassification. The work in [Lalonde et al., 2006] suggests some filters to remove the bad effect such as *isolated surface filter*, *isolated density filter*, ect. Nevertheless, the results just showed mean performance while the required computation is quite expensive. So, we do not use them in our approach. Importantly, we have tested the performances of both [Lalonde et al., 2006] our algorithm and the algorithm in with the CMU dataset (http://datasets.visionbib.com), the classification results confirm that our appproach outperforms the other (note that there exists no ground truth for vegetation detection based on point cloud analysis, this is just another evaluation on another dataset).

8.4 Conclusion

We have presented a new approach to terrain classification based on structure for automobile exploration, in a large variety of object scenarios. The neighbour distance variation inside edgeless areas features clearly discriminates the smooth and scattering areas which are presumably to denote artificial constructions and vegetation, respectively. The conditional local point statistics features complement to a more

8. TERRAIN CLASSIFICATION BASED ON STRUCTURE

complex object classification based on 3D point distribution. In addition, the use of Efficient Graph-based technique for 3D point cloud segmentation in advance helps in avoiding the edge effect affecting to the above features' properties. Consequently, our approach brings a significant improvement for terrain classification.

Chapter 9

A Novel Approach of Terrain Classification for Outdoor Automobile Navigation

The investigation of reconstructing 3D model of the viewed scene has showed good performance in environments such as in yard, hall way or on road. However, in cluttered outdoor environments where frequently the scenes are unknown and the objects are no more static and rigid, the only use of 3D-point analysis is not sufficient to give good decision for safe navigation. Therefore, we on the other hand address a new approach which reconstructs completely 3D scene based on calibrating Laser Scanner and CMOS camera and doing segmentation to result objects in form of region of interest. As a result, the characteristics of each region are then expressed through their corresponding feature vectors, including 2D and 3D features. This is the first time the term of feature vector used to describe a 3D object respecting to the analysis of 3D-point clouds given by a LiDAR. Finally, we also prove that the proposed approach leads to more robust and faster processing and decision-making in terrain classification compared with conventional approaches or pixel-based approaches.

This work has been published in **Proceedings of IEEE International Conference on Computer Science and Automation Engineering** [Nguyen *et al.*, 2011a].

9. A NOVEL APPROACH OF TERRAIN CLASSIFICATION

9.1 Introduction

According to the literature of autonomous navigation for outdoor mobile robot, the main task is to get from point A to point B. While this sounds to be rather simple that the vehicle just follows the Global Positioning System (GPS) breadcrumbs, it is actually a huge problem. In real-world applications, the vehicle has to deal with a variety of terrain, obstacle avoidance, roll-over stability and much more. Even though, the research on the field of autonomous navigation has been started for more than a decade years, giving a methodology which can achieve a full safe autonomous navigation system in outdoor environment is still a daunting challenge. Indeed, there are many publications respecting to this field where we can find a dozen approaches and ways to evaluate the risk and give solution to the vehicle during his operation. Most of methods up to date only stop at comparing the global map planing and local map planing to guide the decision-making of the robot. The deep interpretation of the current surrounding is still ignored, which usually leads to some miss decisions or lost ways. Therefore, this study pays more attention on how to understand the surrounding of the robot and also keeps in mind the given tasks that the robot has to carry out. For that aim, we have built an autonomous mobile outdoor robot(AMOR) which is equipped a laser scanner SICK-LMS221 and CMOS camera: Logitech ProCam 9000, mounted at the front. Actually, we also use other accessories like compass for direction and skewness information, ultra sound sonic sensor for reflectance, and even a helicopter for surrounding and location information (see **Fig. 2.2** in section 2.1). However, we just discuss here the use of laser scanner and CMOS camera in describing surroundings in this work. We tried to result objects in form of regions of interest (ROI) by segmenting the depth image established from scaling distance information given by a LiDAR to greyscale. Each region now represents one unknown object.

The segmentation technique applied in this work is so called Graph-Cut technique. The basic idea of this technique is to compare the internal difference and component difference between two neighbor regions. The concrete *difference* here in this paper is the distance difference to the robot. The initial work of [Felzenszwalb

9. A Novel Approach of Terrain Classification

& Huttenlocher, 2004] inspired the technique to segment colour images regarding the colour difference in three channels: Red, Green and Blue. The segmentation results are impressive for those standard images, regarding to white balance and standard light condition, but shows mean performance for irregular images. Fortunately, the distances given by the ladar SICK LMS221 are very precise which leads to a nice depth image which can be well segmented by the Graph-Cut technique. Secondly, unlike K-mean and other common segmentation techniques, Graph-Cut technique segments image into regions based on both local and global properties as well as object position taken into account. This satisfies the goal of extracting objects. We will prove that the segmentation results are not either too coarse or too fine. Alternatively, it is the first time feature-based approach is presented in this paper to classify terrain, regarding to 3D-point cloud analysis. Previous approaches are based on neighbour or local pixels' consideration, which are time-consuming and less robustness.

In contrast, the feature approach potentially initiates a more robust terrain classification system under real-time constraint (indeed, the LMS221 can get a whole frame of scanning within a half second). The key idea of the approach is to fuse 2D and 3D features extracted from each object or ROI to create discriminative feature vectors. So, a coarse calibration needs to be done to interact 2D and 3D scenes. The coarse calibration is implemented respecting to the time-consuming and robustness, because a full-calibration between the two vision systems will cost much computation and time but show mean performance in outdoor environment [Leidheiser, 2009]. Our calibration method is experimentally proved fast and efficient to the purpose of mapping 2D-3D information.

The next contribution of the paper is to introduce 2D-3D feature fusion which leads to more discriminative feature vectors. While, 2D and 3D features are extracted independently for CMOS image and 3D-point cloud respectively, the interaction will be taken place by mapping them into corresponding objects or ROI to generate a feature vector. Finally, Support Vector Machine is presented in order to train and test the feature vectors. The comparison of the proposed method and other conventional methods is also discussed.

The structure of the paper is organized as following: Related Works in section

9. A NOVEL APPROACH OF TERRAIN CLASSIFICATION

9.2, 2D-3D Coarse Calibration in section 9.3, Feature-based Classification in section 9.4, Experiments and Results in Section 9.5, and Conclusion in section 9.6.

9.2 Related Works

The early approaches to terrain classification were based on the physical properties to provide semantic descriptions of the physical nature of a given terrain region. These descriptions can be associated with nominal numerical physic parameters or traversability estimates to improve traversability prediction accuracy Halatci *et al.* [2007][Manduchi, 2005][Iagnemma & Dubowsky, 2002]Lalonde *et al.* [2006]. Specifically, oriented to the mobility capabilities of an AMOR, there are some other approaches relied on terrain parameter identification via wheel-terrain interaction analysis and terrain classification based on auditory wheel-terrain contact signatures [Iagnemma & Dubowsky, 2002]. However, a large variety of terrain exist together with scenes are often near monochromatic that makes the classification become more challenging. In order to overcome the problem, some researchers suggested to combine the traversability parameters and obstacles detection-based parameters [Halatci *et al.*, 2007][Iagnemma & Dubowsky, 2002]. For that aim, one or more CCD/CMOS cameras are mounted at front of the robot. The 2D cameras are positioned to look down to see the front terrain. One of the highlight benefits is to utilize colour information to detect lethal obstacles at front. The approaches are very abstractive and efficient for applications of planetary exploration rovers like Mars exploration. The restriction of those approaches is just focusing on estimating the traversability of terrain by its physical properties and potential obstacle estimates, while safe autonomous navigation requires more knowledge of surrounding, especially on cluttered environment such as cornfield and off-road. One of the quite early approaches to describe the surrounding of the robot is presented by [Lalonde *et al.*, 2006]. The most remarkable contribution of this approach is to classify surrounding into three classes: *scatter* represents porous volume objects such as tree canopy and tall grasses, *linear structure* denotes to thin objects like wires, thin branches,

9. A Novel Approach of Terrain Classification

and small tree trunk, *surface* describes ground surface, rocks and large tree trunk. In order to do so, a searching cube (with changeable size but often 10x10x10 cm^3) slides around in space to select local points. The statistic features of the selected points are then extracted based on their 3D distribution (please see more details in [Nguyen *et al.*, 2010a]). The disadvantage of this way is how to choose the suitable size of the cube. The size should be changed in term of distances, complexity of the viewed scene, etc. Especially, it turns out worst in case of presence of dense edges which was explained clearly in our previous work [Nguyen *et al.*, 2010b]. The previous work also gave a solution for eliminating *edge effect* by segmenting 3D-point cloud in regions of homogeneous distances. In general, such approach shows good performance in detecting *surface* objects but mean performance in distinguishing *linear structure* and *scatter* objects. Besides, this approach is still a pixel-based terrain classification one which just stops at classifying scenes into three classes due to local point distribution analysis. From our perspective, we think that this is the time to approach a higher level of terrain classification whereby the robot would know where is grass, tree, wall, road, and other obstacles specifically from the viewed scene. This is also the most motivation of this paper.

9.3 2D/3D Coarse Calibration

The calibration between Laser scanner and CCD/CMOS camera becomes very important due to growing need of robust object detection and recognition applications in outdoor environment. However, a robust calibration is time-consuming and computational expensive as well as the performance of the calibration drops significantly in outdoor environment containing vegetation [Leidheiser, 2009]. Two of major reasons are the light intensity and light colour changes and vibration of vegetation, which make interested points of both Laser Scanner data and CMOS image unstable. In fact, regarding to the purpose of terrain classification, we do not need a very precise calibration but reasonable one. In our previous work, we introduced a simple 2D/3D mapping method to project the image plane of a 2D sensor to the 3D coor-

9. A NOVEL APPROACH OF TERRAIN CLASSIFICATION

dinate of a LiDAR, see section 4.2. In fact, the method provided visual pleasing but small objects were often mis-reconstructed. This would not affect to the result of detecting vegetation which commonly appeared as a large object. However, for the aim of classifying different object (possible in small or intermediate size) types in this work, the simple mapping is not precise enough. Hence, we prose a 2D/3D coarse calibration which is relatively robust while still being simple and fast.

Logitech Webcam Pro 9000 and LMS221 are used similar to the work in [Nguyen et al., 2011b]. The applicable distance range of this coupled system lies in the interval $[3.8 \div 15.8][m]$ when the Laser Scanner is positioned $5cm$ higher the CMOS camera. The nearest distance threshold is set due to diminishing the stereo effect [Liu et al., 2008], while the farthest distance threshold is selected due to the quality of data acquisition of the Laser Scanner. The assumption to have all objects located at least 3.8 meters far from the system sounds weak for navigation applications but it is acceptable for our goal of classifying terrain and understanding surrounding of the robot. So, this proposed method is not really well applied for on-board navigation but for prediction and interpretation of the surrounding.

Firstly, a single camera calibration needs to be carried out for CMOS camera to obtain the camera matrix and distortion parameters, which consequently can be used to get undistorted images from raw input images. This step will remove the radial and tangential lens distortion. In this work, we use OpenCV library to do the calibration. From then, whenever we mention CMOS images, we are implicitly talking about the images undistorted.

Secondly, when the Laser Scanner and CMOS camera are positioned very near each other, the views of the two sensors are approximately quite the same in a narrow viewing angle, see **Property 1** in section 4.2. The geometric model of the coupled system can be described as in **Fig. 9.1**. At a distance d, we assume that the sizes of scenes of Laser Scanner scene and CMOS scene are (x_1, y_1) and (x_2, y_2) respectively. Let (α_1, θ_1) and (α_2, θ_2) denote the apertures of Laser Scanner and CMOS camera, respectively. Then, we will have a geometric property as below:

Property 2: *The ratio of sizes of the Laser Scanner scene to of the CMOS scene is equal to the ratio of the tangents of the corresponding aperture angles.*

9. A Novel Approach of Terrain Classification

The **property 2** can be formulated as:

$$\frac{x_1}{x_2} = \frac{tan(\alpha_1)}{tan(\alpha_2)}; \quad \frac{y_1}{y_2} = \frac{tan(\theta_1)}{tan(\theta_2)} \tag{9.1}$$

Thus, the projection of the CMOS scene onto the image plane of CMOS camera will have a size of $(x_1/\delta_x) \times (y_1/\delta_y)$ where (δ_x, δ_y) is the pixel size of a CMOS image. Approximately, the projection of the LS scene on the image plane should have a size of $(x_2/\delta_x) \times (y_2/\delta_y) \equiv (x_1 \frac{tan(\alpha_2)}{tan(\alpha_1)}/\delta_x) \times (y_1 \frac{tan(\theta_2)}{tan(\theta_1)}/\delta_y)$ because we are assuming the views of the coupled system are the same. Similarly, the projection of the intersection part of both scenes (see **Fig. 9.1**) on the image plane should have a size of $L_s = (x_1/\delta_x) \times (y_2/\delta_y)$ or:

$$L_s = x_2 \frac{tan(\alpha_1)}{tan(\alpha_2)}/\delta_x \times y_2/\delta_y \tag{9.2}$$

Specifically, in this work, the size of CMOS images is 640, so the size of projected LS scene is $(640\frac{tan(41)}{55}) \times (480\frac{tan(77)}{70}) \equiv 390$ and the size of the projection of the intersection part is 390. In order to do the projection, we are going to build a grid plane of $(x_1/\delta_x) \times (y_1/\delta_y)$ pixels which will **store the distance values of points** from LS scene projected to the image plane. Of course, the number of points are much

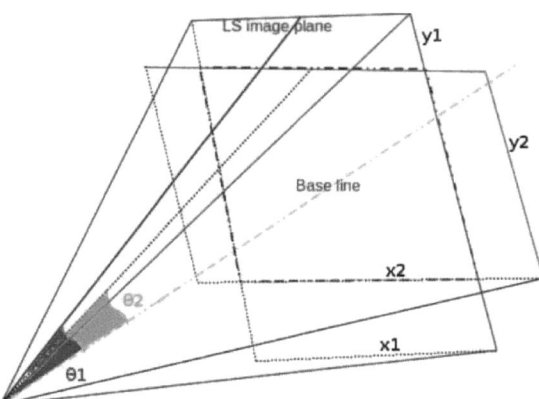

Figure 9.1: Geometric model of Laser Scanner and CMOS scene planes.

9. A NOVEL APPROACH OF TERRAIN CLASSIFICATION

lesser than the number of pixels so that we have to put each point into a suitable pixel in the grid disorderly based on their relative positions. For example, for each scanned line, we only have *nSPL* (number of scanned points per line) scanned points and *nPPL* (number of pixels per line) pixels on the corresponding line in the grid. The first and last point correspond to the first and last pixel (see **Fig. 9.2**). The pixel step in x-coordinate can be calculated as:

$$\lambda_x = \frac{x_{nSPL} - x_1}{nSPL} \qquad (9.3)$$

So the ith scanned point will be stored in the kth column with:

$$k = Round(\frac{x_i - x_1}{\lambda_x} \frac{nPPL}{nSPL}). \qquad (9.4)$$

Where (x_i, y_i, z_i) denotes the position of the ith scanned point.

Similarly, for each column of scanned data, the first and last point correspond to the first and last pixel. Let *numSL* is number of scanned lines, and *nPL* is number of pixel lines. The pixel step in z-coordinate can be calculated as:

$$\lambda_z = \frac{z_{numSL} - z_1}{numSL} \qquad (9.5)$$

So the ith scanned point will be stored in the jth row with:

$$j = Round(\frac{z_i - z_1}{\lambda_z} \frac{nPL}{numSL}). \qquad (9.6)$$

Hence, we have a grid plane containing points projected from Laser Scanner

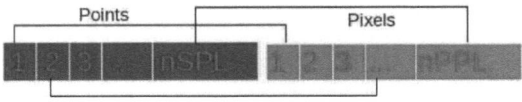

Figure 9.2: Putting points from LS scene onto the grid plane per line.

9. A Novel Approach of Terrain Classification

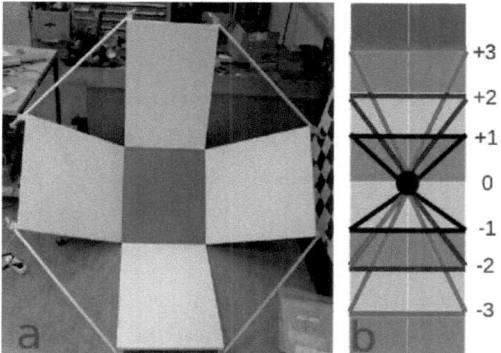

Figure 9.3: a) 3D chessboard model for Laser Scanner and CMOS camera calibration [Leidheiser, 2009]. b) Sketching planes from the centre of the searching window in different levels.

scene and empty pixels. A linear interpolation process will help us to fulfil the empty pixels. The outcome is an image which approximately represents the projection of Laser Scanner scene onto the image plane of CMOS camera, so called projected image. Ideally, the outcome image now can be mapped directly with the CMOS image to reconstruct 3D scene if the pinholes of both devices are positioned at one point as in the geometric model in **Fig. 9.1**. Nevertheless, the Laser Scanner is positioned $7cm$ above the CMOS camera so that there is a shift between the two images. In order to figure out the shift, we have built a 3D chessboard model as shown in **Fig. 9.3**. The four corners of the orange centre box are detected by both Laser Scanner and CMOS camera, which are also displayed on both CMOS image and projected image. Assume that the coordinate of each corner in CMOS image is (x_i, y_i), in projected image is (X_i, Y_i), $i = 1 \div 4$. In an ideal case, the shift between the two images can be calculated as: $shift_x = X_i - x_i$ and $shift_y = Y_i - y_i$ where $X_i - x_i = X_j - x_j$ and $Y_i - y_i = Y_j - y_j$: $\forall (i,j) = 1 \div 4$. In practical, we obtain the mean value of them expressed as:

$$shift_x = \frac{\sum_{i=1}^{4} X_i - x_i}{4}; \quad shift_y = \frac{\sum_{i=1}^{4} Y_i - y_i}{4} \quad (9.7)$$

9. A NOVEL APPROACH OF TERRAIN CLASSIFICATION

Actually, the computation of the shift can be carried out with only one corner. However, the sampling frequencies of Laser Scanner and CMOS camera are totally different, so the value of each scanned point from Laser Scanner projected on the image plane is not absolute but relative. Thus, the average of the shifts calculated via each corner practically gives more robust result. Moreover, the standard deviation of them inspires the belief of accurateness where the less standard deviation there is, the greater the precision of calibration.

Finally, we obtain three parameters of the coarse calibration including $L_s, shift_x$, and $shift_y$, which are fixed for a specific set-up of the coupled system. Even though,

Figure 9.4: Examples of calibration results.

9. A Novel Approach of Terrain Classification

this is a coarse calibration method, its performance is very impressive. The maximum error of the calibration results is less than 5 pixels for all objects located at 3.6m or more far from the coupled system. In fact, the calibration just needs to be done off-line once, the returned parameters can be used from then on until the configuration of the coupled system is changed. As a result, the processing time for an on-line 3D reconstruction is at around a decade of milliseconds with Pentium IV, CPU $2GHz$ and $2GB$ of RAM. Examples of 3D reconstruction results are illustrated in **Fig. 9.4**.

9.4 Feature-based Classification

The traditional terrain classification methods focus on analysing local point distribution and measuring neighbour point relation in order to obtain discriminative properties of points which can classify the points into different classes. These methods can just only be used to query the traversability of the viewed terrain while the aim of object detection and recognition is infeasible. As introduced in the section I, this paper will present a feature-based approach to classify terrain and surrounding objects. The new approach enables a robust terrain classification and make the task of object detection and recognition feasible.

Let us define the term of Image of Interest (IoI) and Depth Image of Interest (DIoI) as below:

Definition 1: The Image of Interest is the image cropped from the CMOS image with size of L_s and the same image centre.

Definition 2: The Depth Image of Interest is the image cropped from the projected image with size of L_s and the image centre shifted amount of $(shift_x, shift_y)$.

Therefore, the IoI and DIoI images respectively represent the colour and distance information of the intersection part of the Laser Scanner scene and CMOS scene. The following subsection will show how to result objects in form of region of interest by segmenting the depth image.

9.4. FEATURE-BASED CLASSIFICATION

9.4.1 Depth Image Segmentation

Unlike pixel-based approaches, the feature-based one has to detect objects first before extracting discriminative features. Thus, an image segmentation needs to be done. Nevertheless, if the segmentation is implemented on the colour image or IoI, the affection from light conditions and colour changes significantly degrades the results, especially with the presence of shadow areas. In addition, the appearance of shadow is inevitable in outdoor environments. Consequently, we come up with the decision of segmenting the depth image or DIoI. Fortunately, the distances given Laser Scanner are very precise whereby we can obtain a very fine depth image which is quite stable even under complex conditions and environments. The segmentation technique used in this work is so called Efficient-Graph-Cut which is firstly introduced by [Felzenszwalb & Huttenlocher, 2004].

Figure 9.5: Examples of segmentation results.

9.4. Feature-based Classification

The technique was used to segment a colour image based on selecting edges from a graph, where each pixel corresponds to a node in the graph, and certain neighbouring pixels are connected by undirected edges. The weights of edges are calculated by colour distances between pixels and also adjusted by the degree of variability in neighbouring regions of the image. The joint-decision of two neighbour regions are made if the maximum distance of two arbitrary pixels in each region is superior or equal the weight of edge of the two regions (see more details in [Felzenszwalb & Huttenlocher, 2004]). If in the work of [Felzenszwalb & Huttenlocher, 2004] the distances of pixels are the differences of colour information, in our work they are the differences of distance information. The reason we use the Graph-Cut is because this technique considers both local and global properties of the scene and the results are not either too coarse or too fine. In order to implement the Graph-Cut segmentation, we have to set some initial parameters like minimum size of regions S_{min}, σ for Gaussian smoothing applied to input images, and the threshold K to controls the degree to which the difference between two components must be greater than their internal differences in order for there to be evidence of a boundary between them. The practical shows that the σ should be fixed at 0.8 while the minimum size should be $S_{min} = 750$ pixels and the threshold K should be 250 for such number of scanned points from 6437 (= 41x157) to 16810 (41x410) (normal scanning modes). If the number of scanned points of Laser Scanner is higher, then $\sigma = 0.8$, $K = 300$, and $S_{min} = 1000$. **Fig. 9.5** shows certain results of the segmentation.

9.4.2 2D/3D Feature Fusion

The previous section has shown the way to project 3D information onto the colour image. This enables a 2D-3D feature fusion process. While the 2D-features are the features extracted from CMOS images, the 3D-features are extracted from the 3D-point clouds of Laser Scanner. From the subsection A, objects are already resulted in form of regions of interest (ROI) in the depth image or DIoI. The ROI are then projected into the IoI and 3D-point cloud to obtain the corresponding ROI in both CMOS image and 3D-point cloud. Therefore, 2D and 3D features of each ROI can

9.4. FEATURE-BASED CLASSIFICATION

be created separately before gathered to generate a feature vector which describes the characteristics of the object.

9.4.2.1 3D Features

The 3D features can be local point distribution statistic [Lalonde et al., 2006], neighbour distance variation [Nguyen et al., 2010b], tactile [Halatci et al., 2007], statistical distributions of 3D data point [Manduchi, 2005]. If the features help to classify 3D points into several classes in pixel-based approaches, we on the other hand figure out the percentage of each class in each object and consider it as one component of the object's feature vector.

Fitting plane:

Besides, we also present here a new approach to measure the 3D-spatial distribution of local points of 3D-point cloud through statistic method. Generally, 3D data is stored as a matrix where each scanned line from Laser Scanner corresponds to a row of the matrix. A window is slided across the matrix to select local points, with one pixel at each shifting unit from the left to the right, from the top to the bottom. The main idea is to build a *fitting plane* so that the summation of point-to-plane distances from selected points is minimum. The local point distribution property of each selected point is then represented by the value of the summation, so called local distribution weight of the point. An update of the weight of each point is taken place only if its new weight is smaller than the current one.

Assume that, there are M local points selected. In order to get a scale invariant feature vector later on, a local coordinate normalization process applies to all local points to normalize the values of each point's coordinate.

$$x_i = x_i - \bar{x}; \quad y_i = y_i - \bar{y}; \quad z_i = z_i - \bar{z} \tag{9.8}$$

Where $\bar{u} = \frac{\sum_i u_i}{M}$ subject to u = {x,y,z}. (x_i, y_i, z_i) and $(\bar{x}, \bar{y}, \bar{z})$ are the ith point's coordinate and the average coordinate of the local points, respectively.

Assume that $W_0 = \{w_0[i] : i = 1 : M\}$ is a set of the current weights of the M local points. The new weights can be computed by minimizing the summation of point-

9.4. Feature-based Classification

to-plane distances from the points to a plane : $ax+by+cz=0$, as in **Eq. 9.9** and **Eq. 9.10**. In other words, we have to build a *fitting plane* whereby the summation of distances from the points to the plane is minimum.

$$w = \underset{\{a,b,c,d\}\in \Re;(a\times b\times c\times d)\neq 0}{argmin(dst(a,b,c,d))} \quad (9.9)$$

Where

$$dst(a,b,c,d) = \frac{\sum_{i=1}^{M}|ax_i+by_i+cz_i+d|}{\sqrt{a^2+b^2+c^2}} \quad (9.10)$$

This turns out to be an optimization problem containing absolute values. The accurate solution is so complex and not always feasible, thus, we come up with a coarse solution implemented by the following algorithm.

Algorithm:

Initiation: $w_o[i] = \infty$; $\forall i = \{1:M\}$.

- Assume that the centre of the current window is $C(x_c, y_c)$. Sketch a set of planes given by the centre C, the left $L_k(x_{c-1}, y_{c-k})$ and the right $R_k(x_{c+1}, y_{c-k})$ subject to $k = \{-3:3\}$ (see **Fig. 4b**).

- Calculate *dst* for each plane.

- *w* is set as the minimum value of *dst*.

    for $i = 0$ to M **do**
        if $w < w_o[i]$ **then**
            $w_o = w$;
        else
            do nothing
        end if
    end for

- Shift the window one pixel and repeat.

189

9.4. FEATURE-BASED CLASSIFICATION

We propose new 3D features, so called *geometric similarity features*, being formed by two components: the *Mean Value* and *Standard Deviation* of the weights of the 3D-points projected in each region of interest (see **Eq. 9.11**). Even though the proposed method introduces a coarse solution, we will prove that the result is robust enough and very impressive, especially for wall fence, concrete road and building detection.

$$\mu_{gsf}^{(k)} = \sum_{i=1}^{N} \frac{w[i]}{N}; \quad \sigma_{gsf}^{(k)} = \sqrt{\sum_{i=1}^{N} \frac{(w[i] - \mu^{(k)})^2}{N}} \quad (9.11)$$

Where N: the number of 3D-points projected into the kth region of interest.

9.4.2.2 2D Features

The 2D features can be histogram distances, colour descriptors and textures. Firstly, histogram distance features are very efficient in detecting unknown-shape, homogeneous colour objects such as vegetation and road, but show mean performance in recognizing multiple-colour objects such as human and arbitrary obstacles. Indeed, our previous work proved that even though there are changes of light intensity and light colour, a very distinctive colour feature can still be obtained for vegetation detection by measuring histogram quadratic distance with the quantization proportion set is **20:4:3** corresponding to Hue:Saturation:Value in HSV colour space [Nguyen et al., 2011b].

Histogram Quadratic:

$$d_q = sqrt(\frac{1}{M}(H_k - H_v)^T * A * (H_k - H_v)) \quad (9.12)$$

$$A_{ij} = \frac{|H_v[i] - H_v[j]|}{max_{m,n}(H_v[m] - H_v[n])} \quad (9.13)$$

Where M is the number of histogram bins. H_v is histogram model (of vegetation,

9.4. Feature-based Classification

wall fence, road or etc). H_k is the histogram of a query object, which has to be normalized: $\sum_i H_k[i] = \sum_i H_v[i]$. A is the cross correlation matrix of histogram bins of H_v. So, A can be computed beforehand to reduce the on-line computation (please see more [Nguyen et al., 2011b]).

In this work, we also prove that the histogram feature can be well applied to detect homogeneous colour objects like concrete roads and wall fences. In order to detect unknown colour objects like human with different cloths or multiple colour obstacles, texture and colour descriptors should be taken into account. In fact, a structured overview given by [van de Sande et al., 2010] has described and compared the invariance properties and the distinctiveness of colour descriptors applied for image retrieval. Similarly, we can apply those invariant colour descriptors for terrain classification where the descriptors are extracted for each ROI instead of the whole image. Although, [van de Sande et al., 2010] has proved that three descriptors including Transformed Colour, Moment Invariants, and Transformed Colour Scale-Invariant Feature Transform (SIFT) are invariant against light intensity and light colour changes, the applicability of those descriptors in object recognition is quite poor. The reason is because they are so strong features that require really precisely matched patterns in order to judge if the query object belong to a class or other class. For instance, even one object is captured from different view angles or different light condition, its representations are usually judged to belong to different classes. The out-performance is given by Opponent SIFT and RGB SIFT, where the Opponent SIFT is a SIFT descriptor in opponent colour space defined as following (see more in [van de Sande et al., 2010]).

$$\begin{pmatrix} O_1 \\ O_2 \\ O_3 \end{pmatrix} = \begin{pmatrix} \frac{R-G}{\sqrt{2}} \\ \frac{R+G-2B}{\sqrt{6}} \\ \frac{R+G+B}{\sqrt{3}} \end{pmatrix} \qquad (9.14)$$

Finally, the texture feature used in this work is Haar WaveLet of Gabor Filter (*HWoGF*) which was firstly introduced by [Nguyen et al., 2010a]. The feature captures texture properties of objects from different angle views and different object scales, thus, the feature is invariant under rotation and scaling.

We have introduced many features in the this section, including both 2D and 3D features. Actually, regarding to the processing time, we would not use all but some specific features for a particular object detection, which will be explained in details next section.

9.5 Experiments and Results

We applied the proposed approach on the same database as in our previous work [Nguyen *et al.*, 2011b], where 500 scenes of cluttered outdoor environments are captured by both Laser Scanner LMS221 and CMOS camera Logitech QuickCam Pro 9000, in both morning and afternoon condition. 300 pairs of 3-D point clouds and CMOS images were used for training and the other 200 pairs were used for testing. The maximum distance set for the Laser Scanner is at about 16 meters. In fact, the idea of combining 2D and 3D features is a quit merit approach, however, the influences of those features affect to the classification result are really difficult to be interpreted or in other words the problem of training the features is a non-linear one. Proposed by [Cortes & Vapnik, 1995] SVM is one of the best available machine learning methods which can deal with non-linear problems. Indeed, our previous work has shown good performance with 81.49% of precision in detecting vegetation using SVM [Nguyen *et al.*, 2011b]. Those motivations drive us to once again use SVM in this work. Nevertheless, there are more objects need to be detected, so we use Multi-classes SVM and One-against-all SVM instead of binary SVM. **Fig. 9.6** and **Fig. 9.7** show the average precision of the proposed method for object detection where the classification results were evaluated by comparing the output of classifier with the hand-labelled data. To obtain the labelled data, colour images are firstly segmented into regions of interest (ROIs). The hand-labelled data contains all ROIs (segmented regions), which are manually labelled as human, tree trunk, road, vegetation, building, and sky. Multi-class SVM helps to classify objects simultaneously where seven features are used, including HWoGF, Histogram Intersection,

9. A Novel Approach of Terrain Classification

Histogram quadratic distance, RGB SIFT, Opponent SIFT, Local point statistic, and Geometric similarity. This is time-consuming because the computation for extracting HWoGF and Histogram quadratic distance features is very expensive. The consuming time of the whole evaluation process is at about one second. In reality, we do not need to detect all but some of them, so we also introduce the single object detection by using One-against-all SVM. In this case, we tried to figure out which features characterize a particular object, thus, the number of features is reduced. For example, in order to detect vegetation, practical experiments show that three features including Geometric similarity(Red),Histogram quadratic distance (Black) and Local point statistic (Pink) are most useful and discriminate (see **Fig. 9.7**).

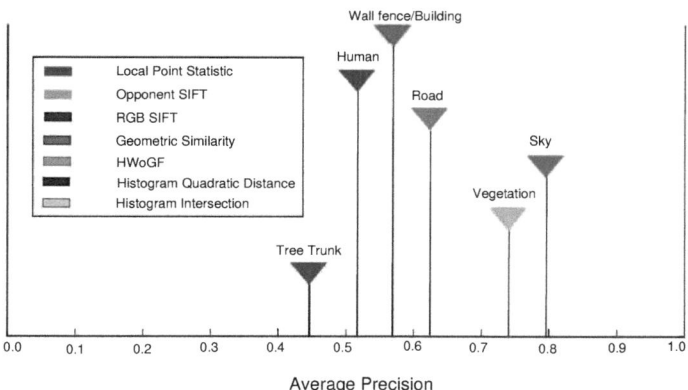

Figure 9.6: Examples of classification evaluation (in percentage) when applied Multi-classes SVM where seven features are used.

Except the data acquisition process of Laser Scanner is very slow, at about two seconds, the other processes are quite reasonable when using One-against-SVM (see **Table 9.1**).

The fastest speed of scanning reached is at about half second using a sweeping reflectance mirror [1] [Schlemper *et al.*, 2011] . However, regarding to the robustness

---
[1] The laser scanner is kept fixed while an additional rotating mirror is used to reflect received signals to the laser's

9. A NOVEL APPROACH OF TERRAIN CLASSIFICATION

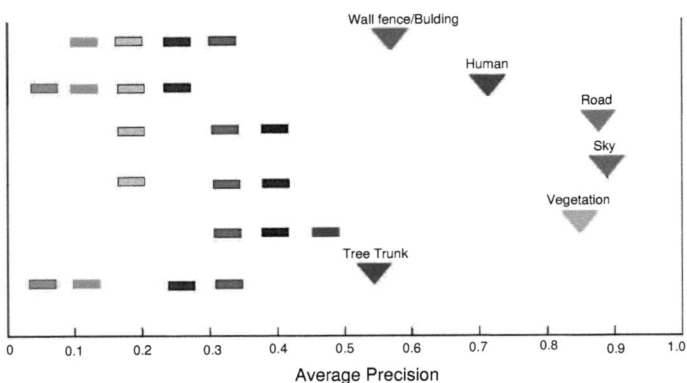

Figure 9.7: Examples of classification evaluation (in percentage) when applied One-against-all SVM where some specific features are used to detect a particular object. *Note: Road is concrete and we also use elevation information in order to detect roads.*

Table 9.1: Precision and Times

Types	Precision	Times(ms)
Tree trunk	0.543	280
Human	0.725	402
Wall/Building	0.586	223
Sky	0.882	435
Vegetation	0.864	593
Concrete road	0.875	435

and stability of current system, we do not apply the technique in this paper but might be in the near future work.

---

eye. The speed of scanning now is proportional to the speed of rotating the reflectance mirror.

9.6 Conclusion

We have introduced a novel approach of terrain classification for outdoor automobile navigation. The approach reveals a very high precision of terrain classification and helps the automobile to understand surroundings completely. This results many benefits in obstacles avoidance, object localization, local and global map planning. The farthest applicable distance is at about sixteen meters, so it is far enough for initializing an up-coming scene prediction application. Indeed, a robust terrain classification system is approximately reached, however time issue is still a challenge for on-board automobile navigation. Therefore, this approach is up to now just applied for scene prediction and surrounding interpretation. Regarding to navigation applications, the method is used to deal with some tough situations that an automobile has to deal with, like blocked around by tall grasses in a cornfield or stopped at a corner in forest environments. Actually, we are building a new way to speed up the data acquisition process of Laser Scanner, that hopefully can improve the reliability of the proposed approach in the near future.

9. A NOVEL APPROACH OF TERRAIN CLASSIFICATION

Chapter 10
Conclusions

10.1 Summary

The thesis has addressed the problem of understanding vehicle environments through vegetation detection and terrain classification tasks, which are explained to be at the core of any control system for efficient autonomous navigation in outdoor environments. We identified the weak points of the existing methods to be inability to cope with complex representation of vegetation, lighting changes, and sharp transition in belief distribution. Since roughness of beliefs is caused mainly by range discontinuities; various appearances of vegetation come from either variety of vegetation species or lighting changes, it is important to develop methods capable of handling those real world phenomena. To this end, we proposed five novel approaches for vegetation detection and two novel approaches for terrain classification. Additionally, we conducted a fitting plane algorithm for depth correction in stereo imaging, which is useful in case stereo cameras are applied to collect the world information instead of a LiDAR. To evaluate these novel approaches, we also implemented other remarkable approaches in the field from our best knowledge and thoroughly compared all their performances through a diverse set of databases and real robotics experiments. Overall, these proposed approaches far outperformed existing methods by several orders of magnitude.

With regard to vegetation detection, it is well-known that vegetation needs to ab-

10. CONCLUSIONS

sorb more red and blue light for its photosynthesis process while strongly reflecting NIR light due to the structure of the leaf; this has been successfully exploited in the remote sensing field to detect green areas on the earth surface. We pointed out that a difference of viewpoints between a satellite and a mobile robot explained why it is a problematic thought to apply the available multi-spectral approaches which are unstable in a robotics context with additional complications (shadow, shining, view of sky). Further more, we proposed the use of an active NIR lighting source which enabled a stable multi-spectral system (Chapter 3). By doing regression analysis on the changes of red and NIR reflectance in terms of luminance observed by such stable multi-spectral system, we derived a new vegetation index, the so-called Modification of Normalized Difference Vegetation Index. Practical experiments confirmed that the new index far outperformed other indices and other existing methods regarding vegetation detection in different lighting conditions and under different illumination effects. Still, the limitations of the method were that its performance degraded sharply in dim lighting conditions; it could not help to distinguish between vegetation and strong NIR reflectance or warm objects (a common issue for all multi-spectral approaches).

Secondly, human perception-based approaches were also surveyed and investigated due to the fact human eye without doubt could easily detect vegetation. Concretely, we presented a 2D/3D feature fusion approach (Chapter 4) using a calibrated vision system which contains a CMOS CMOS camera and a LiDAR. Thereby typical visual features extracted from colours (green, yellow,red-orange,brown) and 3D structures (linear, scatter,surface) were deeply exploited and combined to train a robust vegetation classifier. Consequently, this method was able to provide a high accuracy (83.36%) in detecting a variety of vegetation species which might appear in different colours. Nevertheless, the limitation of the approach was that it required a fully scanned 3D scene while the processing time of data acquisition of LiDAR for a good resolution was quite slow (2 seconds for acquiring 6437 3D points). This lowered the speed of an entire sequence. As a result, this approach could not be applied for guiding on-board autonomous navigation but possibly used in tough situations (corner, stuck point, forest path) where the robot must run slowly or stop, and

10. Conclusions

thus the time would not be really critical.

Chapter 5 realised our greedy ambition in combining both human perception-based and multi-spectral approaches while still heading to a final real-time system. For that aim, we suggested the use of an integrated vision system, the so-called MultiCam, which mounts both CMOS sensor and PMD sensor into a molecular setup, and thus able to provide all data needed for both multi-spectral and visual methods, while not breaking the real-time constraint. Impressively, the MultiCam could provide simultaneously colour, NIR intensity and depth information as fast as a regular video camera. In this work, visual features extracted from colour, texture and spatial distribution were fused with vegetation indices to form optimal vector components to generate an optimal feature vector. Consequently, the resulted classifier yielded a detection accuracy of over 95% while the frame rate was up to 2 fps. This enables a robust and real-time vegetation detection system. The fact is that the more features used, the more robustness obtained, but the more computational expensiveness paid. There is always a trade-off between accuracy and time in applying this method, in which a user might give modification to suite his/her purpose. Nonetheless, the performance of this method, as that of other classification-based methods, depends on the dataset and known scenarios. Thus, it is recommended to collect a sufficient training data which contains possible scenarios that the robot might has to deal with in a specific task.

To obtain a fast and efficient vegetation detection, a spreading algorithm was conducted (Chapter 6). Starting by thresholding vegetation indices, chlorophyll-rich vegetation is detected and considered seeds of a spread vegetation. The algorithm relies on two parallel processes, the so-called vision-based spreading and spectral reflectance-based spreading, to extend the spread vegetation. The first process estimates colour and texture dissimilarities between a seed and its neighbours in order to judge whether or not a neighbour is joined. The second process control an overspreading of the first one by restricting possible regions of spreading for each iteration. We have proven that the algorithm is neither too fine nor too coarse. We are able to detect variety of vegetation in different sunshine conditions, even when all shining, shadow, underexposure, overexposure effects as well as view of sky are taken

10. CONCLUSIONS

into account. Furthermore, by dividing dataset into several groups mainly based on light conditions, we show a concrete and detailed performance comparison between all available methods in detecting general vegetation in different situations. Overall, this proposed algorithm outperforms others, and thus is considered as the most efficient and robust vegetation detection mechanism.

To this end of our investigation on vegetation detection, we aim to answer the question of whether or not detected vegetation is passable. Thus, we introduced a novel approach for a double-check of passable vegetation detection (Chapter 7). This novel approach relies on an estimation of compressibility or less-resistance of vegetation, which is realised by assessing the moveability of vegetation effected by wind. Thus, we addressed a system design where blowing devices were used to create wind to effect vegetation. Also we provided an architecture for the passable vegetation detection system, in which moving vegetation was detected from mapping foreground objects (given by a motion detection and compensation) to detected vegetation (given by a multi-spectral approach). The degree of less-resistance of the moving vegetation was then estimated by referencing its recorded movement through an optical flow detection. Finally, moving vegetation with higher degree of less-resistance is more likely to be detected as passable vegetation. For the purpose of guiding robot navigation, we restricted the region of interest right at the front of the robot. Whereby, the affection of illumination and lighting changes would be significantly reduced, and thus we yielded a very high detection accuracy, at about 98%. However, it should be clear that this method has been only verified to be applicable for scenarios where a robot has already stopped due to tall vegetation as an obstacle at the front, and thus tries to check its traversability.

With regard to terrain classification, we addressed a local point statistic analysis method (Chapter 8); wherein instead of sliding a cube, cylinder, or sphere in space to select a local region, we suggested to segment the point cloud using an efficient graph-cut technique. Thereby, the problematic selection of a volume size is not the case. We have proven that the segmentation is neither too coarse nor too fine, and the classification results are improved significantly for variety of terrains. In addition, we introduced a novel spatial feature, the so-called distance variation inside edgeless regions, to capture the smoothness property of objects. Whereby, we are able to classify between rough surface (low grass, thick bushes) and smooth surface (wall, flat ground), which is infeasible in previous works. Still, the given method requires a fully scanned 3D scene, which is time-consuming. Thus, it is usually used as a post-processing method.

Alternatively, we proposed a machine learning-based method for terrain classification (Chapter 9). More precisely, support vector machine techniques (multi-classes and one-against-all) were used to train different object classifiers. Again, a combination between LiDAR data and colour vision information was exploited. Indeed, seven visual features were extracted from both 2D and 3D information, including local point statistic, opponent SIFT, RGB SIFT, geometric similarity, HWoGF, HQ, and HI. In the first experiment, we applied multi-classes SVM to train the multi-classes classifier which is able to detect six different objects simultaneously, including tree trunk, human, wall/building, road, vegetation, and sky. Consequently, the detection rate was high for vegetation and sky but quite low for others. In the second experiment, we used one-against-all SVM. Different sets of visual features were trained and tested to optimise the feature vector components in order to detect different objects individually. With regard to the accuracy, the second method performed better than the first one. However, one classifier could only allow to detect a single object. Thus, we might need to train many classifiers in order to detect many objects. Overall, it is not clear which one is better, thus one might choose a method which suits one's purpose.

10.2 Discussion

While a variety of approaches have been proposed for both vegetation detection and terrain classification, it would be great to discuss why many approaches are given and what are the advantages and disadvantages of each. Since perceptual inference is very challenging in outdoor environments where exist much uncertainty together with illumination changes, which leads to the infeasibility of a general, complete so-

10. CONCLUSIONS

lution for any perception task. Therefore different approaches exploit the way they handle the trade-off of accuracy, reliability, and efficiency in different sketched scenarios. Indeed, the investigation on point cloud analysis leads to model 3D structures of surrounding objects and flat-ground plane, which are worthy and efficiently used for aiding autonomous navigation of an AGV in highly structured environments. Current autonomous navigation systems still rely mainly on this traditional approach because the range information given by a LIDAR is precise and stable, while other imaging sensors are strongly sensitive to illumination changes in outdoor environments. Yet, the task of interpreting the point cloud is very difficult and not always feasible due to the range discontinuity and the complexity of environments. Hence, even though robotic research has been deeply studied for decade years with regard to outdoor autonomous navigation, there is still room to improve the accuracy of this very early approach. Actually, the conditional local point statistic analysis introduced in Chapter 8 improves the detection accuracy of 10 % for linear structure (wire, branch of tree), and $10 \rightarrow 17$ % for scatter (needle tree, canopy) and surface (wall, ground) structures compared with previous approaches which solely relied on 3D point distribution. When the aim is not only to point out traversable and non-traversable areas of the terrain but also to classify it into object types, e.g. ground, pedestrian, vehicle, and vegetation, it is not feasible to rely solely on LIDAR data which is not informative enough. Meanwhile, images contain abundant information about objects, and thus 2D/3D feature fusion is naturally derived when taking more colour and texture information into account for a better understanding of objects/classes/instances. In general, this approach shows better performance in either terrain classification or vegetation detection compared with the sole LIDAR based analysis, but also requires more time for 2D/3D calibration and 2D/3D feature fusion, please reference Table 8.1 and Table 9.1. Still with around 2 fps, 2D/3D feature fusion approaches are preferred to be only used in tough situations, e.g. at the corner, stuck points where all paths are blocked by dense geometric obstacles, to enable a possible solution. In contrast, for on-board navigation, LIDAR data is solely used for modelling the ground surface and detecting regions of interest (ROI) of possible obstacles while lethal obstacles are then detected by analysing their shape, colour

10. Conclusions

and texture inside the projected ROI on the image plane. Overall, it is clear that depending on different purposes, applications, and situations, different approaches (or preferably called as different strategies) should be applied.

Exceptionally, vegetation is studied in depth and separately from other objects due to its diversity of species existed in a variety of appearances and shapes. While vegetation detection is trivial by human visual system, it arises certainly the idea of imitating human vision in terms of interpreting vegetation appearance. In addition, a coupled system (LIDAR and CMOS/CCD camera) is currently essential for autonomous navigation, which provides both 2D and 3D information of the viewed scene. Therefore investigating colour, texture and 3D spatial distribution is intuitively a good direction to go, the so called human vision (or also 2D/3D feature fusion) based vegetation detection, with no additional device required for the AGV. Since previous approaches only aimed to detect green vegetation, Chapter 4 proposed green/red-orange/yellow colour models together with local point statistic analysis of point cloud, that lead to detect efficiently vegetation in its three common colours. The accuracy was obtained at about 82.86% for 500 outdoor scenes captured in different illumination conditions. Because this is the first investigation on fusing 2D/3D features to detect different colour vegetation, so there exists no previous ground truth for comparison. However, if compared with previous approaches, which analyse solely LIDAR data or colour image to detect green vegetation, the proposed method shows much better performance in accuracy, please reference Table [4.3, 9.1] and [Zafarifar & de With, 2008]. Still one might concerns about the slow speed of this approach, at about 2 fps. Again, a good argument is that when the aim is just to guide an AGV mainly for autonomous on-road driving, the on-board navigation relies on the ground model given by point cloud analysis, the 2D/3D feature fusion approach is only applied when getting stuck or uncertainty occurred. In such situation, the time is not very critical. When the goal is to navigate most of the time in vegetated terrains, this approach is no more applicable. For off-road driving, it is advisable to use light spectral reflectance approaches to detect vegetation, as presented in Chapter [3, 5, 6]. While Chapter 3 relies solely on the light spectral reflectance property of vegetation, Chapter [5 & 6] try to improve the accuracy by taking into account more

10. CONCLUSIONS

visual features. The comparison in performance of different approaches for general vegetation detection was concretely reported in Table [3.2, 5.1, 6.1]. Whereby the spreading algorithm in Chapter 6 outperforms the others when taken into account the complexity of scenarios, illumination effects and real-time constraint. Due to the fact that the idea of coming up with the spreading algorithm was derived from the investigation on vegetation indices as well as colour and texture analysis in Chapter [3 & 5], respectively. Hence, the three approaches are presented in this thesis for a better understanding. In addition, the work described in Chapter 7 demonstrates that when considering the specific task as passable vegetation detection for autonomous navigation guidance of an AGV, the vision system is specifically configured to restrict the view right at the front of the AGV, whereby vegetation indices show more advantages than the spreading algorithm. Indeed, with the restricted view the affection of illumination effects is significantly reduced, and thus the performances of the approaches are similar while the vegetation indices are much faster. Again, this confirms that while there exists no ideal solution for vegetation detection in cluttered outdoor environments, the choice depends significantly on the purpose of applications and the sketched scenarios for experiments. In a general case, the spreading algorithm shows the best performance among all available vegetation detection approaches.

10.3 Direction for Future Work

Classification-based methods for vegetation detection or terrain classification (see Chapter [4, 5, 9]) are very straightforward and intuitive where human perspectives of colour vision and 3D spatial distribution are exploited to model objects. The accuracies of such approaches are rather high when many features are used to train the object classifier. The fact is that the more features are used, the more robustness is obtained, but the more computational expensiveness are paid. Thus, there is always a trade-off between precision and processing time when applying those approaches. This restricts the applicability of the classification-based methods for on-board navigation applications. Indeed, such a vegetation detection or terrain clas-

10. Conclusions

sification module using those techniques is just turned on when dealing with a very tough situation like at the corner, at the stuck point, or in a forest path where the robot is running very slowly, and thus time is not critical.

There are two possibilities for the future work to do in order to improve the performance of those approaches. First, regarding a software development, finding optimal feature vector components to train vegetation classifier might decrease the processing time while still keeping the high accuracy of the approach. Second, regarding a hardware development, even though the suggestion on using a rotating mirror instead of sweeping up and down the 2D Laser Scanner helps to improve four times the data acquisition speed, it is still not fast enough for many real-time applications. Clearly, the PMD camera is much faster than the LiDAR in acquiring distance information, it can not operate properly in outdoor environments due to the huge affection from the sunlight. Thus, an investigation on improving the speed of LiDAR in data acquisition or an innovation in depth imaging of Time-of-Flight sensors in outdoor is really appreciated.

In line with the same idea of including information from different types of sensors to improve the object detection/classification and allow the classification of multiple class of objects, there is another possibility or direction to investigate, the so-called high-level fusion framework. Instead of finding a set of features from multi-sensors' data, then training them with machine learning techniques to generate a multi-class classifier (see Chapter [4, 5, 9]), a set of object detectors/classifiers might be applied, then the corresponding classification evidences might be fused by a generic high-level sensor fusion framework using probability theories. Thus, the advantages and drawbacks of an object detector/classifier can be complemented with the others to improve the overall performance of such the object detection/classification. Let's start with a simple example. Assume that we have two classification hypotheses provided by a LiDAR target detector, and four classification hypotheses (pedestrian, bike, car, truck) provided by a camera target detector. When a big object is detected by the LiDAR, we would expect the object should be car or truck. The probability that the object might be a car is $P_{L-car} = 50\%$, a truck is $P_{L-truck} = 50\%$, a bike is $P_{L-bike} = 0\%$, a pedestrian is $P_{L-pedes} = 0\%$. Classification evidences (detected

10. CONCLUSIONS

from several continuous frames) provided by the camera target detector can be used to estimate the probability that the object might be a car is P_{C-car}, a truck is $P_{C-truck}$, a bike is P_{C-bike}, a pedestrian is $P_{C-pedes}$. Finally, we are able to estimate the probability that the object, for instance, might be a car as $P_{car} = P_{L-car} \times P_{C-car}$. Although this generic high-level sensor fusion framework has not been deeply investigated, the idea is really promising. The future work should provide a concrete comparison between this framework and the feature fusion approaches presented in Chapter [4, 5, 9].

With regard to the light spectral reflectance approaches, it is quite naturally that the idea arises to use several active lighting sources in different light spectral ranges to compensate deficiencies of object representations in a narrow light band, and thus improve the result of detecting vegetation or even other objects. By observing representations of different objects in different light spectral bands, we might be able to obtain discriminative characteristics of those objects. It should be clear that we are not promoting the traditional multi-spectral approach which needs a long time spectral scanning process, and the required hardware is very expensive; meanwhile the performance varies significantly depending on different sunshine conditions. We on the other hand encourage the method using several active light bands such as red, blue, green, and infrared, so that we are able to modulate the emitted light in order to receive selective light. Also we can adjust the intensities of received light, thus reducing the affection from the sunlight.

Indeed, multi-spectral approaches using an active lighting source have been proven to be accurate and efficient in detecting general vegetation (see Chapter [3, 6, 7]). Nevertheless, when there is a significant change of background brightness, we still need to manually adjust the exposure times of the two sensors in the MultiCam, in order to achieve optimal performance. This limitation mainly comes from hardware issues that those sensors have low dynamic ranges, thus being too sensitive against lighting changes. Therefore, it is advisable to build a new multi-spectral system which includes two high dynamic range cameras and an active NIR system, in order to have a more stable system. We suppose here three possible setups for that new system.

10. Conclusions

Figure 10.1: Monocular setup for the new multi-spectral system.

In **Fig. 10.1**, the new multi-spectral system can be established by replacing the two sensors in the MultiCam with two high dynamic range sensors. This setup requires a professional design for the whole system and technology to integrate all components in one compact device.

Roughly, one might build a similar system which realises the multi-spectral system by combining two high dynamic range cameras, cold mirror and LED NIR lighting system, as seen in **Fig. 10.2**. It is also possible to position the two cameras as in a stereo setup, see **Fig. 10.3**. This is much simple but we have to calibrate these cameras as well as a stereo effect is inevitable.

10. CONCLUSIONS

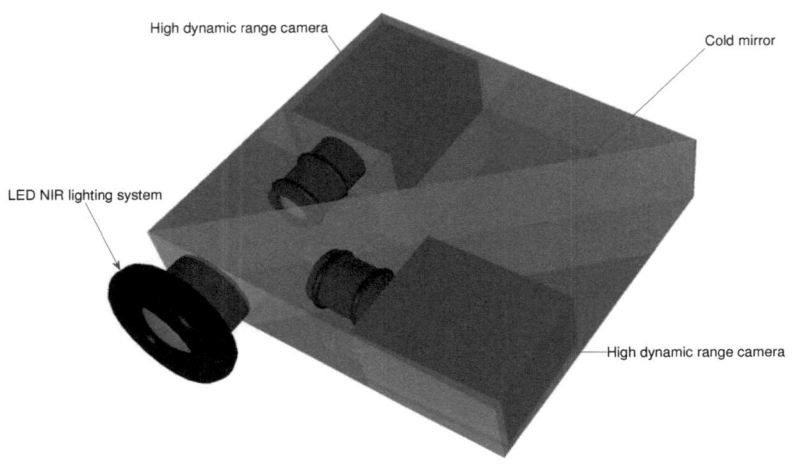

Figure 10.2: Rough monocular setup for the new multi-spectral system.

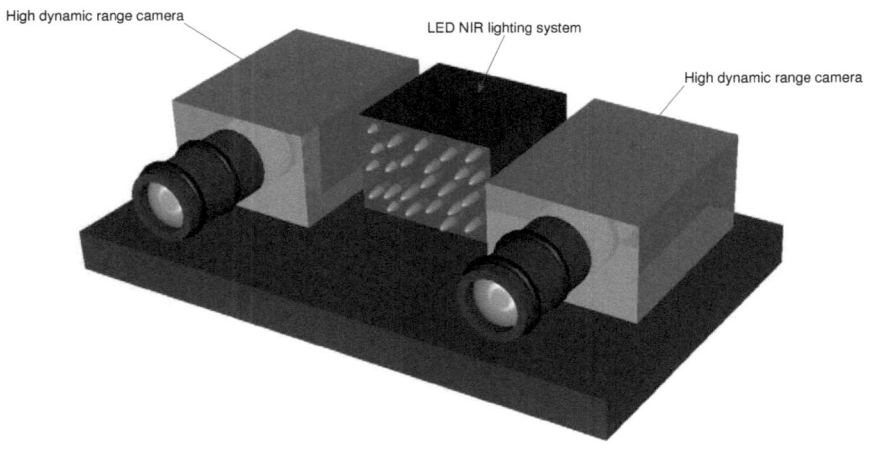

Figure 10.3: Stereo setup for the new multi-spectral system.

Appendix A - Expert Concerns and Rebuttal

Vegetation Indices Applied for Vegetation Detection (Chapter 3)

Question 1

The Kinect sensor, which is currently under evaluation in many labs around the world, has similar features than the Zess-Multicam. It would be therefore very interesting to know whether the presented approach can be applied on the Kinect as well or not.

Answer

In principle, it seems to be no problem to apply the proposed method with Kinect sensor. However, Kinect sensor shoots out infrared light (please see kinect-shoots[1]). This means that the sensor would receive the scattered light, thus, we would get an NIR image that contains many separate near-infrared points. Therefore we might need region growing techniques to connect those points (it is hard also). We actually are testing the applicability of the proposed method with Kinect sensor. One more thing we concern about is the intensity of received near-infrared light which is rather weak. Zess-Multicam gives a chance to change the intensity of emitting near-infrared light, thus, expectedly adjust the intensity of the received near-infrared light this provides a better stability of the outcome or performance.

---

[1] http://www.extremetech.com/extreme/83908-microsoft-kinect-shoots-out-infrared-light-video

10. CONCLUSIONS

Question 2
How do the authors determine the "level of consideration", a number from 1-5, for the categories "Complex environment" and "Complex illumination". What would be, e.g., a level 3 environment complexity? What is the difference between a level 2 and a level 1 illumination complexity?

Answer
Level of consideration is determined by Taxonomy Classification. Concretely, we considered 5 considerations for illumination complexity, including: intensity-colour change, shadow, shining, underexposure, overexposure. We consider how many effects taken into account from those available approaches. The number of effects considered reveals the level of consideration. For complex environments: level 1: hall-way/in yard/campus; level 2: rough road; level 3: off-road with low-grasses; level 4: off-road with tall-grass/ bushes, level 5: forest.

2D/3D Feature Fusion (Chapter 4)
Question 3
The calibration range of the laser-camera-mapping is 3.8m to 15.8m. During navigation, most interesting things happen in front of the robot ($< 4m$). Why is the mapping still useful for navigation?

Answer
This range depends heavily on the speed of the robot which is already at 3m/s also for navigation, thus, a larger range is necessary than 4m. Moreover, the knowledge about the front scene can help to understand the surrounding, thus, to have a better decision-making in some tough situations. For example, the robot is stuck at the corner of the road inside the forest dues to a wrong odometry (potentially caused by slippage when driving over grass). The knowledge about the front scene together with the global map helped to solve the problem (note: GPS points were limited according to the game rule, and GPS were frequently lost inside the forest).

Question 4
You give the impression that laser scanner data acquisition is time consuming: but SICK works at 10Hz easily with RS232, it can be faster with a serial faster than

10. Conclusions

RS232, and 180 distances per acquisition is a lot less than 640x480 RGB pixels. So, your processing algorithms may be slow, but I do not think SICK data acquisition in itself is slow. Maybe you are referring to simulating a 3D Laser by moving a 2D one. Then, yes, as you have to tilt the SICK to generate the 3D point cloud, it takes time. Is this what you mean? some 3D cameras are appearing in the market. You are using a 3D camera in this role yourself. Although the current 3D cameras have limited performance outdoor, maybe in the future you could cheaply take a 3D cloud at 20 or 30 Hz with commercial sensors.

Answer

Yes, LMS221 needs to sweep up and down to scan 3D points, then we need to put those points into Cartesian Coordinate. Such operations take time to acquire the whole 3D point Cloud (in our case, it takes nearly two seconds for acquiring 6437 points putting them into Cartesian Coordinate). In fact, Velodyne LIDAR is fast and robust but quite expensive, so we are also in the hope that the cheap ones will appear soon.

General Vegetation Detection using an Integrated Vision System (Chapter 5)

Question 5

If you see only a small part of the scene, it does not hamper navigation?

Answer

The narrow angle of view and low image resolution are the main limitation of PMD camera. However, the distances of interest lie in the interval (4,7) meters rather than (0,4) meters as usual (please see the system setup). With that distance, the local scene viewed is not too much small. The reason to focus on the distances (4,7) meters because we want to detect vegetation on the robot's way while the robot is running at around 1m/s to 3 m/s. At that speed, we also did not expect the robot to turn sharply, thus, using the camera does not seriously hamper the navigation. Of course, we still acknowledge that limitation.

Question 6

Why use a low-resolution PMD-prototype when the range measurements cannot be used?

10. CONCLUSIONS

Answer

The multi-spectral camera proposed by [Bradley et al., 2007] suffers strong affection from illumination changes. We use the MultiCam which is equipped with an integrated active lighting system for obtaining a more stable multi-spectral system. Please see the video in youtube-vegetation-detection[2]. Intuitively, the detection is quite stable under different lighting conditions. Additionally, we are developers of PMD-prototype, and we are trying to use different lighting sources (with different wavelengths) simultaneously for the aim of range measurement in outdoor environments, the results are promising. So, we have some reasons to stay with PMD.

Spreading Algorithm for Efficient Vegetation Detection (Chapter 6)
Question 7

The presented algorithm essentially fuses information of three different sources: vegetation indices, colour information, and texture information. While convincing results could be achieved in the experiments, the algorithm seems to be rather ad-hoc. It is based on thresholded pixel seeds that are expanded if neighbouring pixels fulfil certain criteria. To me it seems worthwhile to investigate a probabilistic combination of the different classifiers instead.

Answer

Regarding to using probabilistic combination of the different classifiers. In an early approach, we already tried to use a Markov Random Field (MRF) to model the visual difference (colour and texture) instead of a convex combination as introduced in this paper. However, the trained MRF only helps to detect vegetation which has simultaneously high probabilities of both colour and texture similarity: $MRF = P_{texture}P_{colour}$. Thus, the algorithm could not detect vegetation in a dark region where there seems to be no texture detected: $P_{texture} = 0$. Two vegetation neighbour pixels could not be joined if their colours are to too much different: $P_{colour} = 0$. This hinders the purpose of detecting a variety of vegetation appeared in many different colours. Concretely, vegetation marked inside the red ellipses in the below image, for instance, could not be detected because those regions are too dark, thus,

---
[2]http://www.youtube.com/watch?v=tnVnSDMnl-g

10. Conclusions

Left: Colour image. **Right**: Vegetation marked as green from the proposed algorithm.

no texture found there. The multiplication in MRF degrades the performance of the algorithm in case one feature missed. This drives us to the idea of using a convex combination.

The convex combination between colour and texture similarities help to vote for candidates which dominate texture similarity or colour similarity, or both of them. Certainly, we acknowledge that based on such convex combination the algorithm is rather greedy. However, we already used the spectral reflectance-based spreading to control that.

Regarding to the THRESHOLD of vegetation indices

Vegetation detection is well done by thresholding vegetation indices (Normalized Different Vegetation Index, Perpendicular Vegetation Index, etc.) in a good lighting condition, which is well known in the remote sensing field. However, a frequent occurrence is the presence of illumination effects including shadow, shining, under-exposure over-exposure when operating outdoor. This degrades the performance of vegetation indices significantly due to the changes of all objects' spectral reflectance distribution.

For example, in the left image in the below figure, with a low threshold of vegetation indices, there appears many false negatives. When increasing the threshold, many vegetation areas are not detected, as shown in the middle image. The point is that even in a "bad" light condition with the presence of above effects, we can still

10. CONCLUSIONS

Left: The threshold is too low. **Middle**: the threshold is too high. **Right**: output from the algorithm.

detect several parts of chlorophyll-rich vegetation by setting a high threshold for vegetation indices, see the middle image. Consider detected vegetation as seeds, we can spread them out based on visual features. This is what we aim for. Actually, based on the seeds in the middle image, we can spread them out to obtain the right image using the proposed algorithm. Based on the spectral reflectance distributions of different materials, we can define the "high threshold" for vegetation indices, which helps to detect chlorophyll-rich vegetation in different lighting conditions, please reference [Bradley et al., 2007]. Therefore, the thresholds are not obtained by hand-tuned but from the study of spectral reflectance distributions on different materials.

Regarding to criteria

The criteria is built from assessing the difference of vegetation pixel and its neighbours respecting to colour and texture dissimilarities. First, although different species of vegetation can have different colours, considering a small region of it, the colour is expected to be homogeneous. Second, texture of vegetation is turbulent (this is trivial as you mentioned, but the way of interpreting the turbulent texture is novel in this work, please see the explanation in the third Answer). Based on those properties, a region-growing, graph-cut, etc. techniques can help to expand vegetation. On the other hand, look at the spectral reflectance distribution of vegetation, we expect that MNDVI and NDVI of vegetation pixel should be higher than a lower bound threshold. Then we use the lower bound threshold to restrict possible vegetation regions in the scene, whereby the vegetation expanding based on visual features (colour and texture) should lie inside those regions. Therefore, the algorithm is neither too coarse

10. Conclusions

or too fine.

Question 8

The proposed algorithm was evaluated using specialized equipment (a "MultiCam"). The results will be hard to verify for someone that does not have access to this sensor. Would it be possible to conduct experiments using of-the-shelf components?

Answer

As pointed out in the section 6.5, the work does not strongly require the use of a MultiCam, a multi-spectral camera together with an additional NIR lighting system is suitable. Even in case of not using an active lighting source, the algorithm can still work with NDVI solely. Concretely, we have tested our algorithm from solely thresholding NDVI to generate vegetation seeds. We used stereo cameras with one camera covered by a NIR-Transmiting filter and the other covered by a NIR-Blocking filter (we used those filters from Hoya company http://www.hoyaoptics.com). Alternatively, one might use the setup built as in [Bradley et al., 2007]. The point is that without using the NIR lighting system, vegetation indices could not provide stable results in detecting vegetation due to spectral reflectance changes of all objects in different lighting conditions. Especially in a very dark background as at http://www.youtube.com/watch?v=OWWay2I9Q-E. Thresholding NDVI is just to detect dark areas in the viewed scene (the disadvantages of using NDVI is explained in details in our previous publication [Nguyen et al., 2012c]). In fact, this paper points out that the approach using multi-spectral camera with an additional lighting system added enables a possibility to detect a variety of vegetation in really different lighting conditions. We acknowledge that there exists some limitations from using the MultiCam:

- The resolution is very low for NIR image: 64x48 pixels.
- CMOS sensor has a low dynamic range.

We encourage researchers to built a system which consist of a multi-spectral camera (built as the above figure or using stereo camerasit is better to use High Dynamic Range (HDR) cameras) with an additional NIR lighting system (can use NIR LED for example). The use of an independent lighting source really leads to a stable and robust system. The reason that we use the MultiCam currently is because we are a

10. CONCLUSIONS

producer of PMD camera and MultiCam, and we want to test the possibility of using a multi-spectral camera with an additional lighting system to detect robustly variety of vegetation in different lighting conditions. We have already had a plan to built a system: multi-spectral camera (2 HDR cameras) and LED NIR lighting source, in a future work.

Question 9

How does this spreading algorithm handle false negatives like unwanted holes and bleeding into false positives

Answer

There are two cases:

- Unwanted holes are selected as seeds. This is not usually the case because we already set high thresholds for MNDVI & NDVI to select chlorophyll-rich vegetation. Exceptionally, it is true that a dark and quite hot object can cause a problem (for example, black parts of car/motobike car strongly shined by the sun). In this case, the algorithm does not help to eliminate the object. However, object's pixels could not be expanded out due to high difference in colour and texture dissimilarity measure compared with the neighbouring regions (recall that we over-segment the colour image into many small regions).

- If unwanted holes are not selected as seeds, based on the algorithm. They are only merged into the "spreading vegetation" if they have high MNDVI & NDVI as well as similar colour and texture with neighbouring vegetation. This is not expected to be happened. However in reality, you can see that there exists false negatives in our vegetation detection results, usually at the boundaries of vegetation regions. The problem comes from the low resolution (64x48) and low quality of NIR image, thus, the regions marked as high MNDVI & NDVI are much larger than expected.

References

AMANN, M.C., LESCURE, M., MYLLYLAE, R. & RIOUXM, M. (2001). Laser ranging: a critical review of usual techniques for distance measurement. *Optical Engineering*, **40**, 10–19. 40, 41, 42

ANGELOVA, A., MATTHIES, L., HELMICK, D. & PERONA, P. (2007). Fast terrain classification using variable-length representation for autonomous navigation. In *IEEE Computer Society Conference on Computer Vision and Pattern Recognition*, 1–8, Minneapolis, MN, USA. 144

ANGUELOV, D., TASKARF, B., CHATALBASHEV, V., KOLLER, D., GUPTA, D., HEITZ, G. & NG, A. (2005). Discriminative learning of Markov random fields for segmentation of 3D scan data. In *Proc. IEEE Computer Vision and Pattern Recognition (CVPR)*, 169–176, San Diego, CA, USA. 161

ASNER, G.P. (1998). Biophysical and Biochemical Sources of Variability in Canopy Reflectance. *Remote Sensing of Environment*, **64**, 234–253. xxiv, 66, 67

BAUDAT, G. & ANOUAR, F. (2001). Kernel-Based Methods and Function Approximation. In *International Joint Conference on Neural Networks*, 1244–1249, Washington DC, USA. 98, 171

BERENS, J. & FINLAYSON, G.D. (2000). Log-opponent chromaticity coding of colour space. In *Proc. of the 15th International Conference on Pattern Recognition*, 206–211, Barcelona, Spain. 125

REFERENCES

BILMES, J. (1997). A Gentle Tutorial of the EM Algorithm and its Application to Parameter Estimation for Gaussian Mixture and Hidden Markov Models. Tech. Rep. TR-97-021, Berkeley, CA, USA. 98, 170

BOLEY, D., MAIER, R. & KIM, J. (1989). A Parallel QR Algorithm for the Non-Symmetric Eigenvalue Algorithm. *Computer Physics Communications*, **53**, 61–70. 144

BRADLEY, D.M., UNNIKRISHNAN, R. & BAGNELL, J. (2007). Vegetation Detection for Driving in Complex Environments. In *Proc. IEEE International Conference on Robotics and Automation*, 503–508, Roma, Italy. xxiv, 64, 67, 71, 72, 77, 81, 82, 107, 113, 114, 120, 121, 130, 131, 133, 134, 135, 141, 145, 146, 160, 212, 214, 215

BROX, T., BREGLER, C. & MALIK, J. (2009). Large displacement optical flow. In *In Proceedings of the IEEE conference on computer vision and pattern recognition*, 41–48, Miami, FL. USA. 155

BRUN, X. & GOULETTE, F. (2007). Modeling and Calibration of Coupled Fish-Eye CCD Camera and Laser Range Scanner for Outdoor Environment Reconstruction. In *Proc. Int. Conf. 3-D Digital Imaging and Modeling*, 320–327, Montreal, QC, Canada. 87

CHAKRAVARTI, R. & MENG, X. (2009). A Study of Color Histogram Based Image Retrieval. In *Proc. of the Sixth International Conference on Information Technology: New Generations*, 1323–1328, Las Vegas, NV, USA. 109

CHANG, C.C. & LIN, C.J. (2012). A Library for Support Vector Machines. Http://www.csie.ntu.edu.tw/ cjlin/libsvm. 113

CLARK, R.N., SWAYZE, G.A., LIVO, K.E., KOKALY, R.F., SUTLEY, S.J., DALTON, J.B., MCDOUGAL, R.R. & GENT, C. (2003). Imaging spectroscopy: Earth and planetary remote sensing with the usgs tetracorder and expert systems . *Journal of Geophysical Research*, **108**, 1–5. 69

REFERENCES

COLLINS, E.G. (2008). Vibration-based terrain classification using surface profile input frequency responses. In *IEEE International Conference on Robotics and Automation*, 3276–3283, Pasadena, CA, USA. 144

CORTES, C. & VAPNIK, V. (1995). Support vector networks. Machine Learning. *Journal of Machine Learning*, **20**, 273–297. 98, 113, 170, 192

CRIPPEN, R.E. (1990). Calculating the Vegetation Index Faster. *Journal of Remote Sensing of Environment*, **34**, 71–73. 65, 108, 120

DAHLKAMP, H. (2006). Self-supervised monocular road detection in desert terrain. In *Proc. of Robotics: Science and Systems*, Philadelphia, USA. 87

DUPONT, E.M., ROBERTS, R.G., MOORE, C.A., SELEKWA, M.F. & COLLINS, E.G. (2005). Online terrain classification for mobile robots. In *ASME Conference Proceedings*, 1643–1648, Orlando, FL, USA. 144

DUPONT, E.M., MOORE, C.A., COLLINS, E.G. & COYLE, E. (2008). Frequency response method for terrain classification in autonomous ground vehicles. *Autonomous Robots*, **24**, 337–347. 144

FARNEBÄCK, G. (2003). Two-Frame Motion Estimation Based on Polynomial Expansion. In *Proceedings of the 13th Scandinavian Conference on Image Analysis*, 363–370, Halmstad, Sweden. 155

FECHTELER, P. & EISERT, P. (2008). Adaptive Color Classification for Structured Light Systems. In *IEEE Conference on Computer Vision and Pattern Recognition Workshops*, 1–7, Anchorage, AK, USA. 43

FELZENSZWALB, P.F. & HUTTENLOCHER, D.P. (2004). Efficient Graph-Based Image Segmentation. *International Journal of Computer Vision*, **59**, 167–181. 53, 54, 92, 106, 129, 134, 156, 163, 165, 176, 186, 187

REFERENCES

FINLAYSON, G.D., HORDLEY, S.D., LU, C. & DREW, M.S. (2006). On the Removal of Shadows From Images. *IEEE Trans. Pattern Analysis and Machine Intelligence*, **28**, 59–68. 125

FORSTER, F., RUMMEL, P., LANG, M. & RADIG, B. (2001). The HISCORE camera a real time three dimensional and color camera. In *Proc. IEEE International Conference on Image Processing (ICIP)*, 598–601, Thessaloniki, Greece. 43, 50

FREUND, Y. & SCHPIRE, R.E. (1997). A decision-theoretic generalization of on-line learning and an application to boosting. *Journal of Computer and System Sciences*, **55**, 119–139. 98, 170

GAT, N. (2000). Imaging Spectroscopy Using Tunable Filters: A Review. In *Proc. SPIE Wavelet Applications VII*, vol. 4056, 50–64. 63

GHOBADI, S.E., LOEPPRICH, O.E., LOTTNER, O., AHMADOV, F., HARTMANN, K., WEIHNS, W. & LOFFELD, O. (2008). Analysis of the Personnel Safety in a Man-Machine-Cooperation Using 2D/3D Images. In *Proceedings of the EURON/IARP International Workshop on Robotics for Risky Interventions and Surveillance of the Environment*, 59–66, Benicassim, Spain. 133

GHOBADI, S.E., LOFFELD, O. & RADIG, B. (2010). *Real Time Object Recognition and Tracking Using 2D/3D Images*. Ph.D. thesis, University of Siegen. 43, 46, 103, 133, 147

GRIMSON, W.E.L. (1998). Using Adaptive Tracking to Classify and Monitor Activities in a Site. In *IEEE Conference on Computer Vision and Pattern Recognition*, 22–29, Santa, Barbara, CA. 151

GU, Y.J. & ZHONG, J. (210). Grass Detection Based on Color Features. In *Proc. of CCPR*, 1–5, Chongqing, China. 119

HAFNER, J., SAWHNEY, H.S., EQUITS, W., FLICKNER, M. & NIBLACK, W. (1995). Efficient Color Histogram Indexing for Quadratic Form Distance Func-

REFERENCES

tions. *IEEE Trans. on Pattern Analysis and Machine Intelligence*, **17**, 729–736. 95, 97

HAINDL, M. & ZID, P. (2007). Multimodal Range Image Segmentation. In *Vision Systems: Segmentation and Pattern Recognition (book)*, InTech. 43

HALATCI, I., BROOKS, C.A. & IAGNEMMA, K. (2007). Terrain classification and classifier fusion for planetary exploration rovers. In *IEEE Aerospace Conference*, 1–11, Big Sky, MT, USA. 144, 178, 188

HARITAOGLU, I., HARWOOD, D. & DAVIS, L.S. (1998). W4s: A Real-Time System for Detecting and Tracking People in 2 1/2 D. In *Proc. of the 5th European Conference on Computer Vision*, 877–892. 151

HORN, B.K.P. & SCHUNK, B.G. (1981). Determining optical flow. *Artificial Intelligence*, **17**, 185–203. 155

HUANG, J., LEE, A.B. & MUMFORD, D. (2000). Statistics of Range Images. In *Proc. of IEEE Computer Vision and Pattern Recognition (CVPR)*, 324–331, Hilton Head Island, SC, USA. 161

HUETE, A.R. (1988). A Soil-Adjusted Vegetation Index (SAVI). *Journal of Remote Sensing of Environment*, **25**, 295–309. 72, 145

IAGNEMMA, K. & DUBOWSKY, S. (2002). Terrain estimation for high-speed rough-terrain autonomous vehicle navigation. In *Proceedings of the SPIE Conference on Unmanned Ground Vehicle Technology IV*, vol. 4715, 256–266. 144, 178

JAIN, S. (2003). A survey of Laser Range Finding. Tech. rep. 40, 50

JEONG, S., WON, C.S. & GRAY, R.M. (2004). Image Retrieval Using Color Histograms Generated by Gauss Mixture Vector Quantization. *Journal of Computer Vision and Image Understanding*, **94**, 44–46. 95, 96, 110

JORDAN, C.F. (1969). Derivation of leaf area index quality of light on the forest floor. *Ecology*, **50**, 663–666. 69, 72, 145

REFERENCES

KONG, H., AUDIBERT, J.Y. & PONCE, J. (2010). General Road Detection From a Single Image. *IEEE Trans. Image Processing*, **19**, 2211–2220. 110, 126

KUHNERT, K.D. (2008). Software architecture of the Autonomous Mobile Outdoor Robot AMOR. In *IEEE Intelligent Vehicles Symposium*, 889–894, Eindhoven, Netherlands. 35, 148

KUHNERT, K.D. & SEEMANN, W. (2007). Design and realisation of the highly modular and robust autonomous mobile outdoor robot AMOR. In *The 13th IASTED International Conference on Robotics and Applications*, 464–469, Würzburg, Germany. 35, 148

KUHNERT, L., THAMKE, S., AX, M., NGUYEN, D.V. & KUHNERT, K.D. (2012). Cooperation in heterogeneous groups of autonomous robots. In *Proc. of IEEE International Conference on Mechatronics and Automation*, 1710–1715, Chengdu, China. 35

KUMAR, V.V., RAO, N.G., RAO, A.L.N. & KRISHNA, V.V. (2009). IHBM: Integrated Histogram Bin Matching For Similarity Measures of Color Image Retrieval. *Journal of Signal Processing, Image Processing and Pattern Recognition*, **2**, 109–120. 109

LALONDE, J.F., VANDAPEL, N., HUBER, D.F. & HEBERT, M. (2006). Natural Terrain Classification using Three-Dimensional Ladar Data for Ground Robot Mobility. *Journal of Field Robotics*, **23**, 839–861. 79, 81, 91, 119, 120, 134, 135, 141, 144, 161, 163, 168, 169, 173, 178, 188

LEIDHEISER, J. (2009). *Erzeugung eines texturierten 3D-Modells aus 3D- Tiefendaten und 2D-Bilddaten fr die lokale Kartierung mit dem Auenbereichsroboter AMOR*. Master's thesis, Institute for Real-Time Learning Systems, Univerisity of Siegen. xxv, xxix, 89, 177, 179, 183

LILLESAND, T.M. & KIEFER, R.M. (1987). *Remote sensing and image interpretation*. John Wiley, New York, 1st edn. 71

REFERENCES

LIN, C.J. & CHANG, C.C. (2011). LIBSVM : a library for support vector machines. *ACM Transactions on Intelligent Systems and Technology*, **20**, 1–27. 113

LIU, R., ZHU, Q., XU, X., ZHI, L., XIE, H., YANG, J. & ZHANG, X. (2008). Stereo effect of image converted from planar. *Journal of Information Sciences*, **178**, 2079–2090. 180

LU, L., ORDONEZ, C., COLLINS, E.G. & DUPONT, E.M. (2009). Terrain Surface Classification for Autonomous Ground Vehicles Using a 2D Laser Stripe-Based Structured Light Sensor. In *Proc. IEEE/RSJ. Conf. Intell. Robots Syst.*, 2174–2181, St. Louis, MO, USA. 86, 120, 144

LUAN, X. (2001). *Experimental Investigation of Photonic Mixer Device and Development of TOF 3D Ranging Systems Based on PMD Technology* . Ph.D. thesis, Centre for Sensor Systems, University of Siegen. 41

LUCAS, B.D. & KANADE, T. (1981). An iterative image registration technique with an application to stereo vision. In *Proceedings of Imaging understanding workshop*, 121–130, Vancouver, BC, Canada. 155

MACEDO, J., MANDUCHI, R. & MATTHIES, L. (2000). Laser-based Discrimination of Grass from Obstacles for Autonomous Navigation. In *Proceedings of International Symposium on Experimental Robotics*, 111–120, Hawaii, USA. 144

MANDUCHI, R. (1999). Bayesian fusion of color and texture segmentations. In *Proc. of IEEE Int. Conf. on Computer Vision*, 956–962, Kerkyra, Greece. 86

MANDUCHI, R. (2005). Obstacle Detection and Terrain Classification for Autonomous Off-Road Navigation. *Journal of Autonomous Robots*, **18**, 81–102. 65, 87, 178, 188

MATSUYAMA, T., OHYA, T. & HABE, H. (1999). Background Subtraction for Non-Stationary Scenes. Department of Electronics and Communications, Graduate School of Engineering, Kyoto University: Sakyo, Tech. Report. 151

REFERENCES

MERKLINGER, H.M. (1996). *Focusing the View Camera (book)*. Bedford, Nova Scotia: Seaboard Printing Limited. 40

MÖLLER, T., KRAFT, H., FREY, J., ALBRECHT, M. & LANGE, R. (2005). Robust 3D Measurement with PMD Sensors. Technical Report, PMD-Tech. 44, 45

NAKAI, H. (1995). Non-Parameterized Bayes Decision Method for Moving Object Detection. In *Proc. of the 2nd Asian Conference on Computer Vision*, 447–451, Singapore. 151

NASA (2012). Measuring Vegetation (NDVI & EVI). Available at: http://earthobservatory.nasa.gov/Features/MeasuringVegetation. xxiv, 68

NGUYEN, D.V. (2012). Image databases and example vegetation detection results videos, available at http://duong-nguyen.webs.com/vegetationdetection.htm. 115, 135

NGUYEN, D.V., KUHNERT, L., AX, M. & KUHNERT, K.D. (2010a). Combining Distance and Modulation Information for Detecting Pedestrians in Outdoor Environment using a PMD Camera. In *Proc. of the 11th IASTED International Conference Computer Graphics and Imaging*, 163–171, Innsbruck, Austria. 46, 103, 104, 161, 179, 191

NGUYEN, D.V., KUHNERT, L., SCHLEMPER, J. & KUHNERT, K.D. (2010b). Terrain Classification Based On Structure For Autonomous Navigation in Complex Environments. In *Proc. of the 3th ICCE International Conference on Communications and Electronics*, 163–168, Nha Trang, Vietnam. 79, 81, 91, 92, 93, 120, 144, 160, 161, 168, 179, 188

NGUYEN, D.V., KUHNERT, L., JIANG, T. & KUHNERT, K.D. (2011a). A Novel Approach of Terrain Classification for Outdoor Automobile Navigation. In *Proc. IEEE Int. Conf. Computer Science and Automation Engineering*, 609–616, Shang-Hai, China. 144, 175

REFERENCES

NGUYEN, D.V., KUHNERT, L., JIANG, T., THAMKE, S. & KUHNERT, K.D. (2011b). Vegetation Detection for Outdoor Automobile Guidance. In *Proc. of IEEE International Conference on Industrial Technology*, 358–364, Auburn, AL, USA. 67, 79, 81, 86, 109, 110, 112, 113, 120, 134, 135, 141, 144, 148, 155, 180, 190, 191, 192

NGUYEN, D.V., KUHNERT, L. & KUHNERT, K.D. (2011c). An Integrated Vision System for Vegetation Detection in Autonomous Ground Vehicles. In *Proc. of IASTED International Conference on Robotics*, 447–455, Pittbugh, USA. 80, 81, 102, 103, 114, 145, 148, 155

NGUYEN, D.V., JIANG, T., KUHNERT, L. & KUHNERT, K.D. (2012a). Fitting Plane Algorithm-based Depth Correction for Tyzx DeepSea Stereoscopic Imaging . In *International Conference on Communications and Electronics* , 291–295, Hue, Vietnam. 50

NGUYEN, D.V., KUHNERT, L. & KUHNERT, K.D. (2012b). Spreading Algorithm for Efficient Vegetation Detection. *Journal of Robotics and Autonomous Systems*, **60**, 1498–1507, http://dx.doi.org/10.1016/j.robot.2012.07.022. 118, 146

NGUYEN, D.V., KUHNERT, L. & KUHNERT, K.D. (2012c). Structure overview of vegetation detection. A novel approach for efficient vegetation detection using an active lighting system. *Journal of Robotics and Autonomous Systems*, **60**, 498–508. 69, 106, 120, 121, 122, 123, 130, 131, 133, 134, 135, 141, 145, 147, 148, 155, 215

NGUYEN, D.V., KUHNERT, L., THAMKE, S., SCHLEMPER, J. & KUHNERT, K.D. (2012d). A Novel Approach for A Double-Check of Passable Vegetation Detection in Autonomous Ground Vehicles . In *Proc. 15th IEEE Annual Conference Intelligent Transportation Systems*, 230–236, Anchorage, Alaska, USA. 143

NGUYEN, D.V., KUHNERT, L. & KUHNERT, K.D. (2013). General Vegetation Detection Using An Integrated Vision system . *International Journal of Robotics and Automation*, **28**. 103

REFERENCES

OHTA, J. (2007). *Smart CMOS Image Sensors and Applications (book)*. Crc Press Inc. 50

OJEDA, L., BORENSTEIN, J., WITUS, G. & KARLESEN, R. (2006). Terrain characterization and classification with a mobile robot. *Journal of Field Robotics*, **23**, 103–122. 144

OLIVER, N., ROSARIO, B. & PENTLAND, A. (2000). A Bayesian Computer Vision System for Modeling Human Interactions. *IEEE Trans. Pattern Analysis and Machine Intelligence*, **22**, 831–843. 151

PLAUE, M. (2006). Analysis of the PMD Imaging System. Technical Report, Berlin, Germany. 160

PMD, T. (2009). 3D Video Sensor Array with Active SBI, http://www.pmdtec.com. 50

QI, J., CHEHBOUNI, A., HUETE, A.R. & KERR, Y.H. (1994). A modified soil adjusted vegetation index: MSAVI. *Remote Sensing of Environment*, **48**, 119–126. 72

QUINLAN, J. (1993). *C4.5: programs for machine learning (book)*. Morgan Kaufmann Publishers Inc. San Francisco, CA, USA. 98, 170

RANKIN, A. & MATTHIES, L. (2008). Daytime Mud detection for unmanned ground vehicle autonomous navigation. Tech. rep., Orlando, FL, USA. 144

RASMUSSEN, C. (2001). Laser Range-, Color-, and Texture-Based Classifiers for Segmenting Marginal Roads. In *Proc. of International Conference on Computer Vision and Pattern Recognition Technical Sketches*, Kauai, HI, USA. 87

RASMUSSEN, C. (2002). Combining Laser Range, Color and Texture Cues for Autonomous Road Following. In *IEEE Robotics and Automation (ICRA)*, vol. 04, 4320–4325, IEEE. 161

REFERENCES

RASMUSSEN, C. (2004). Grouping dominant orientations for ill-structured road following. In *Proc. of IEEE International Conference on Computer Vision and Pattern Recognition*, vol. 178, 470–477. 126

RICHARDSON, A.J. & C. L., W. (1977). Distinguishing vegetation from soil background information. *Photogrammetric Engineering and Remote Sensing*, **43**, 1541–1552. 70

ROUSE, J.W., HAAS, R.H., SCHELL, J.A., DEERING, D.W. & HARLAN, J.C. (1974). Monitoring the vernal advancement of natural vegetation. NASA Goddard Space Flight Center, Greenbelt, MD, Final Rep. 70, 145, 148

SABEENIAN, R.S. & PALANISAMY, V. (2009). Texture Based Weed Detection Using Multi Resolution Combined Statistical and Spatial Frequency. *Journal of World Acadeny of Science, Engineering and Technology*, **28**, 549–553. 119

SADHUKHAN, D. & MOORE, C. (2003). Online terrain estimation using internal sensors. In *Proc. of the Florida Conference on Recent Advances in Robotics*, Florida Atlantic University, FL, USA. 144

SAXENA, A., SUN, M. & Y.NG, A. (2009). Make3D: Learning 3D Scene Structure from a Single Still Image. *IEEE Transactions on Pattern Analysis and Machine Intelligence*, **31**. 54, 129

SCHLEMPER, J., KUHNERT, L., AX, M. & KUHNERT, K.D. (2011). Development of a high speed 3D laser measurement system for outdoor robotics. In *Proc. of Eurobot Conference*, vol. 161, 277–287, Prague, Czech Republic. 193

SHULL, C.A. (1929). A spectrophotometric study of reflection of light from leaf surfaces. *Bot. Gazette*, **87**, 583–607. 145

SOBOTTKA, K. (2000). *Analysis of Low-Resolution Range Image Sequences*. Ph.D. thesis, Universität Bern. 48

REFERENCES

SURMANN, H., LINGEMANN, K., NCHTER, A. & HERTZBERG, J. (2001). A 3D laser range finder for autonomous mobile robots. In *Proc. 32nd International Symposium on Robotics (ISR)*. 41, 50

TARPLEY, J.D., SCHNEIDER, S.R. & MONEY, R.L. (1984). Global vegetation indices from the NOAA-7 meteorological satellite. *Journal of Climate Appl. Meteorol*, **23**, 491–494. 65, 70, 120

TOWNSHEND, J.R.G., GOFF, T.E. & TUCKER, C.J. (1985). Multitemporal dimensionality of images of normalized difference vegetation index at continental scales. *IEEE Trans.Geosci. Remote Sens.*, **23**, 888–895. 70, 120

TOYAMA, K., KRUMM, J., BRUMITT, B. & MEYERS, B. (1999). Wallflower: Principles and Practice of Background Maintenance. In *Proc. of International Conference on Computer Vision*, 255–261, Kerkyra, Greece. 151, 155

TUCKER, C.J., FUNG, I.Y., KEELING, C.D. & GAMMON, R.H. (1986). Relationship between atmospheric CO_2 variations and a satellite-derived vegetation index. *Journal of Nature*, **319**, 195–199. 70, 120

TULEY, J., VANDAPEL, N. & HEBERT, M. (2004). Technical report CMU- RI-TR-04-44. Robotics Institute, Carnegie Mellon University. 161

ULLMAN, S. (1979). The Interpretation of Visual Motion. *MIT Press*. 155

ÜNSALAN, C. & BOYER, K.L. (2004). Linearized Vegetation Indices Based on a Formal Statistical Framework. *IEEE Trans. on Geoscience and Remote Sensing*, **42**, 1575–1585. 145

VAN BEEK, J.C.M. & LUKKIEN, J.J. (1996). A parallel algorithm for stereo vision based on correlation. In *Proc. of International Conference on High Performance Computing*, 251–256, Trivandrum, India. 51

REFERENCES

VAN DE SANDE, K.E.A., GEVERS, T. & SNOEK, C.G.M. (2010). Title. *IEEE Trans. Pattern Analysis and Machine Intelligence*, **32**, 1582–1596. 94, 95, 124, 191

VANDAPEL, N., HUBER, D.F., KAPURIA, A. & HEBERT, M. (2004). Natural terrain classification using 3-D ladar data. In *Proc. IEEE Robotics and Automation (ICRA)*, vol. 05, 5117–5122. 91, 163

WELLINGTON, C., COURVILLE, A. & STENTZ, A. (2006). A generative model of terrain for autonomous navigation in vegetation. *Journal of Robotics Research*, **25**, 1287–1304. 87, 144

WILLSTATTER, R. & STOLL, A. (1913). Utersuchungenuber Chlorophyll. 69, 160

WOLF, D.F. & FOX, D.B.W. (2005). Autonomous terrain mapping and classification using hidden markov models. In *Proc. of IEEE Int. Conf. on Robotics and Automation*, 2026–2031, Spain. 87

WOODFILL, J.I., GORDON, G. & BUCK, R. (2004). Tyzx DeepSea High Speed Stereo Vision System. In *Proc. IEEE Computer Vision and Pattern Recognition Worshop (CVPRW)*, 41, Washington DC, USA. 51

WREN, C., AZARBAYEJANI, A., DARRELL, T. & PENTLAND, A. (1997). Pfinder: Real-Time Tracking of the Human Body. *IEEE Transactions on Pattern Analysis and Machine Intelligence*, **19**, 780–785. 151

WU, L., ZHANG, Y., GAO, Y. & ZHANG, Y. (2004). Tree Crown Detection and Delineation in High Resolution RS Image. In *Proc. of IEEE IGARSS*, vol. 60, 3841–3844, Beijing, China. 86

WURM, K., KÜMMERLE, R., STACHNISS, C. & BURGARD, W. (2009). Improving robot navigation in structured outdoor environments by identifying vegetation from laser data. In *Proc. IEEE/RSJ international conference on Intelligent robots and systems*, 1217–1222, St. Louis, MO, USA. 65, 81, 120

REFERENCES

ZAFARIFAR, B. & DE WITH, P.H.N. (2008). Grass Field Detection for TV Picture Quality Enhancement. In *Proc. of International Conference on Consumer Electronics*, 1–2, Las Vegas, NV, USA. 86, 119, 203

ZHANG, G.P. (2000). Neural Networks for Classification: A Survey . **30**, 451–462. 98, 170

i want morebooks!

Buy your books fast and straightforward online - at one of the world's fastest growing online book stores! Environmentally sound due to Print-on-Demand technologies.

Buy your books online at

www.get-morebooks.com

Kaufen Sie Ihre Bücher schnell und unkompliziert online – auf einer der am schnellsten wachsenden Buchhandelsplattformen weltweit! Dank Print-On-Demand umwelt- und ressourcenschonend produziert.

Bücher schneller online kaufen

www.morebooks.de

OmniScriptum Marketing DEU GmbH
Heinrich-Böcking-Str. 6-8
D - 66121 Saarbrücken
Telefax: +49 681 93 81 567-9

info@omniscriptum.de
www.omniscriptum.de

Printed by Books on Demand GmbH, Norderstedt / Germany